© Mark Leech 2017

Publisher: Brueckner Leech

Photographer: All images by Mark Leech unless images credited to other photographers.

Design: Rye Dunsmuir

ISBN: 978-0-9945946-0-0 Paperback

Print on Demand by Ingram Spark www.ingramspark.com

Author

Mark Leech

mleech@iinet.net.au

*Tasman Island from Cape Pillar. Image credit pixeluxe/iStock by Getty Images*

*Sunrise over our farm, Northwest Bay. Image credit Mark Leech*

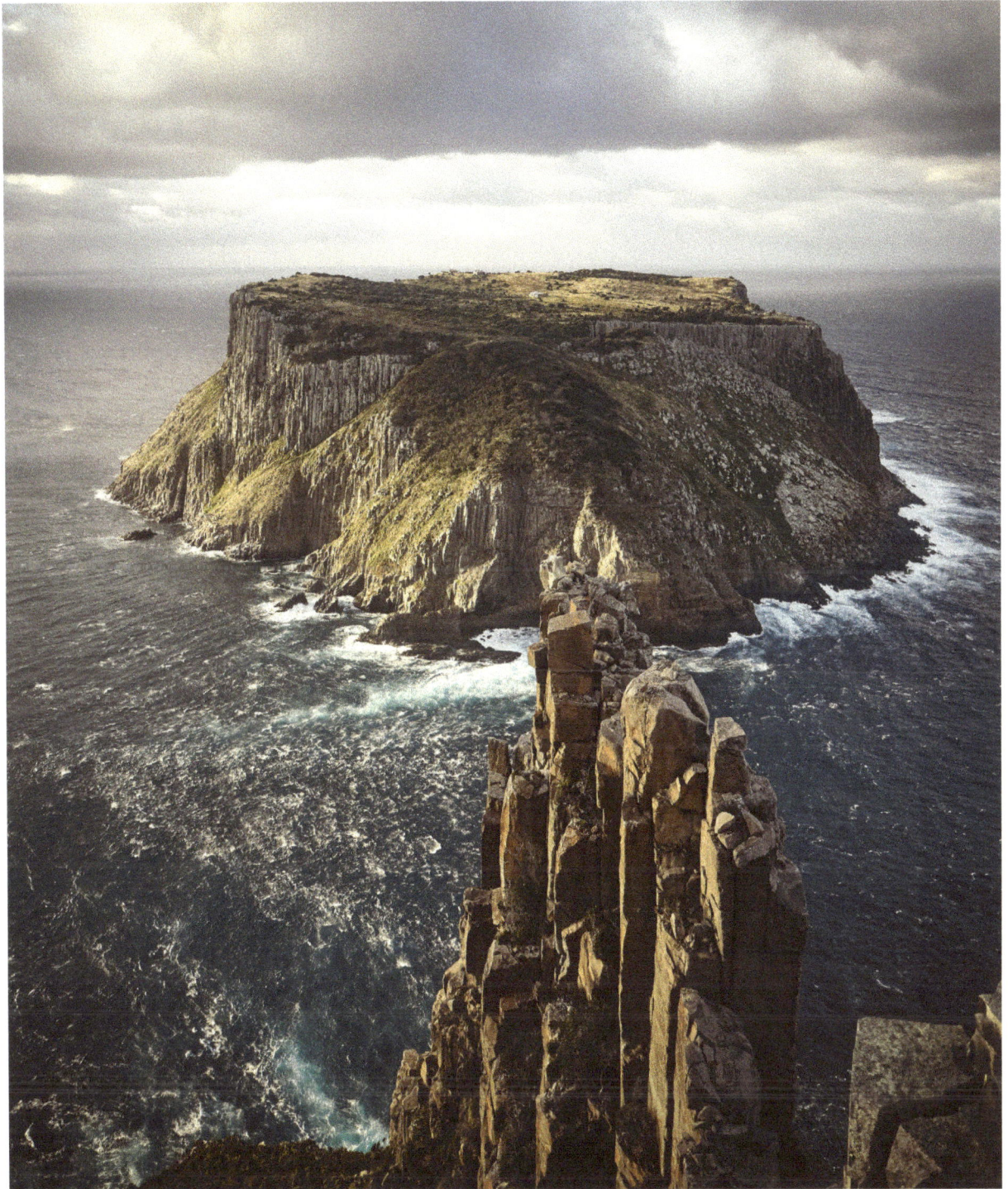

*Tasman Island from Cape Pillar. Image credit pixeluxe/iStock by Getty Images*

*Sunrise over our farm, Northwest Bay. Image credit Mark Leech*

For those of you reading who are not Tasmanians or who have not visited our island I wanted to share a snippet of our history, where we are on the planet and a taste of the islands beauty. To literally taste our world class food you'll just have to come and see!! We welcome you!

First to pay respect to the Tasmanian Aboriginal people, the original inhabitants of the land and the traditional custodians of the island that we all benefit from.

The first European sightings and landing was by the Dutch explorer Abel Tasman in 1642, followed by du Fresne of France in 1772 with quick possession made by the British in 1777. The Empire was expanding and what better way to populate the Antipodes than by creating felons out of minor misdemeanors as well as sending some more hardened folk for company. So began the European settlement of our island.

And the bees were soon to follow, well to be introduced and they thrived.

Initially apiaries as elsewhere were stationary, with abundant flora for spring and early summer but a significant summer dearth. The industry stayed as a small producer until the discovery of our iconic leatherwood, Eucryphia lucida occurring in rainforest from the far northwest to the south. This now provides the backbone of the industry accounting for almost 70% of honey production.

Some further sweeteners; Wall Street Journal in 2014 stated that Tasmania is the next global food destination and Lonely Planet Travel describes Tasmania as one of the must see places of the world, and we get to live here and keep bees. The images show where we are, At The End of the Earth, and just a little of our beauty. The beekeeping journey follows.

*Cradle Mountain and Dove Lake. Image credit TSKB/iStock by Getty Images*

*The Wooden Boat Festival. Image credit Mark Leech*

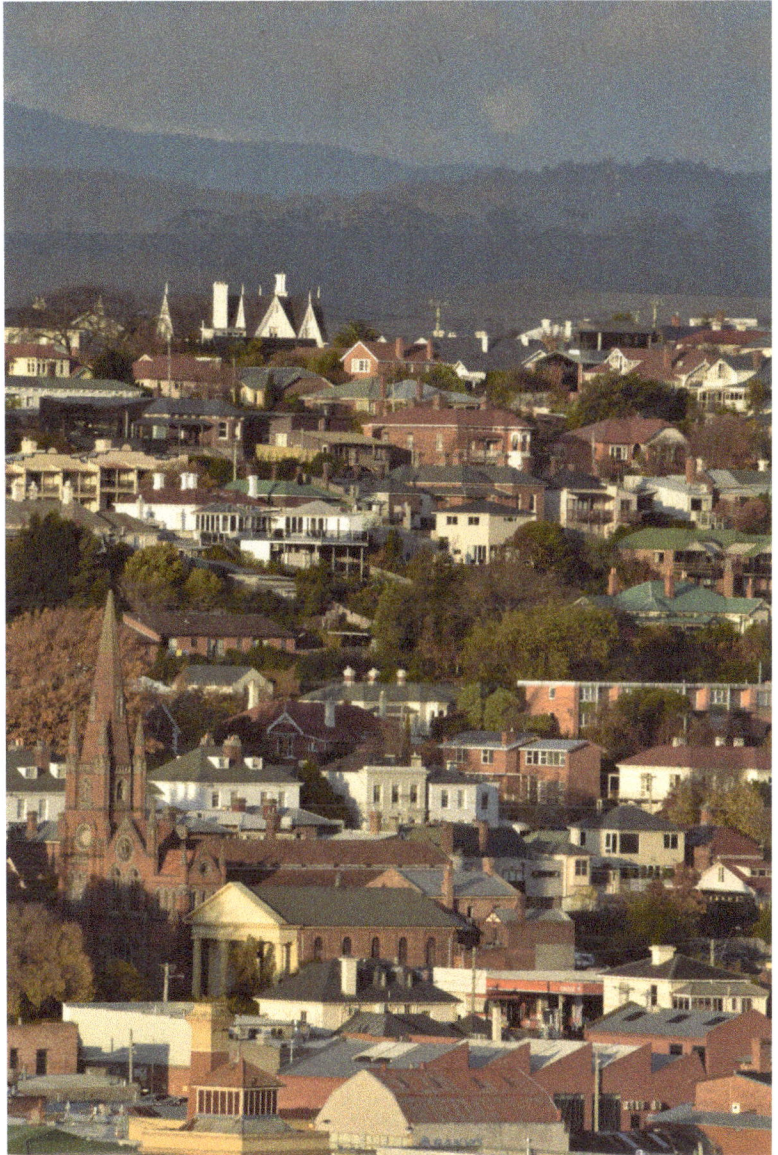

*Launceston scene. Image credit Mark Leech*

*Lavendar Farm at Nabowla. Image credit Tony Feder/iStock by Getty Images*

This book has been a long time coming. I wish to acknowledge the generous support of the Tasmanian Beekeepers Association and the Apiary Industry Disease Committee.

There are many people to thank for their encouragement, technical input as designers, illustrators, photographers, beekeepers, friends and especially family.

The whole book (except for the Chapter on Alternative Beekeeping and Innovation) has been technically edited by the following commercial beekeepers, Lindsay Bourke, Peter Norris, Robbie and Nichola Charles, Ian and Shirley Stephens and Brian Medcraft. I am grateful for their contribution, correction and providing Tasmanian context. The chapter Alternative Beekeeping and Innovation has been generously edited by Ronnie Voigt.

Several beekeepers have provided particular support, direction and help with my beekeeping exploits and writing; Julian Wolfhagen, Lindsay Bourke, Hedley Hoskinson, The Stephens family, Paul Wigger, Yves Ginat, Bruce Direen and especially Laurie Cowen and Ian Hewitt who have always been available, providing their time and help with the book and beekeeping in general.

The book editors had to polish my ordinary English into a more readable form, thank you Steve Graham, Julia Hewitt, Gavin Livingston, my wife Susanne and the technical editors.

My involvement in the apiary industry as an author and researcher has provided many contacts with beekeepers and researchers in Tasmania, the Australian mainland and internationally. This has emphasised how generous beekeepers and the beekeeping community are with their knowledge and time, always willing to share, I am so grateful. A particular thank you to Harold Ayton and former Government Apiary Inspectors for the long running guide, Bee-keeping in Tasmania that has instructed generations of beekeepers. Karla Williams the Government Apiary Bio-security Officer has been a constant help.

Design excellence and inspiration with book layout and design has been generously provided by Rye Dunsmuir of Nology. Mark le Roux thank you for your individualised illustrations and tables.

Photographers from near and far have provided their creative and clear images, Bruce Gibson, Karla Williams, Wellington Apiary, Melbourne City Rooftop Honey, David Barton Photography, Kathy Keatley-Garvey, Zachary Huang, Dan Eisikovich, Des Cannon, Natural Beekeeping Tasmania.

A very special thank you to my wife Susanne for her perseverance, direction, support and for encouraging me to bring the book to completion. To my daughters, sons-in love and beautiful grandchildren for always encouraging and inspiring me to continue. Jon, thank you for being an enthusiastic initiator and passing the beekeeping baton to me.

I thank God for always being there and for the most amazing creature the bee and its community, a super organism, forever inspired.

**CHAPTERS**

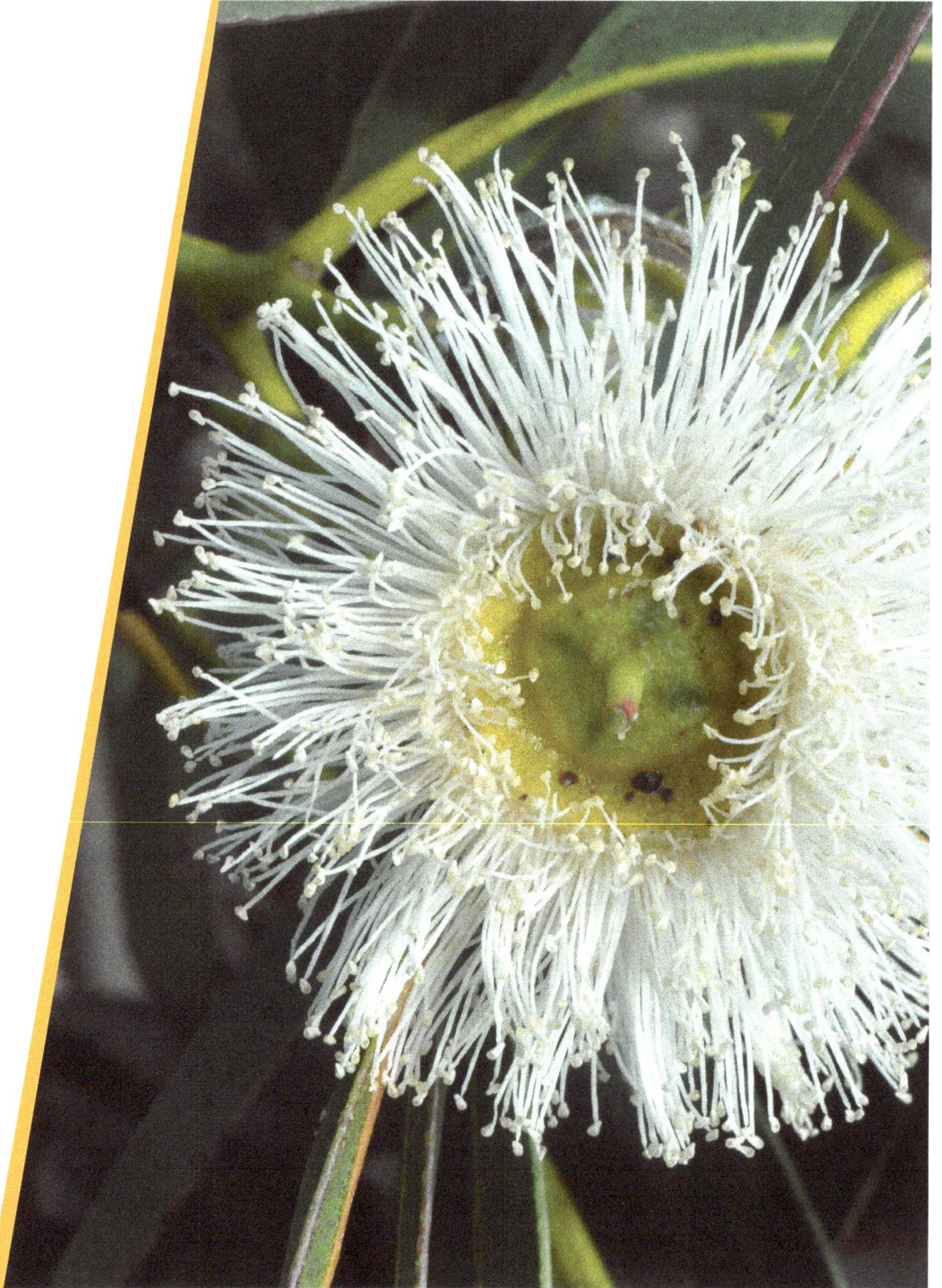

One hundred beekeepers, one hundred different ways to keep bees.

Welcome to the most interesting and captivating interest, hobby, sideline or business you can become involved in. It is a lifetime of learning, as you discover the world of the honeybee and how to be a good steward of your colonies health. Whether you aspire to a stationary hive in your garden or to optimise your honey collecting by following peak floral activity, you will become captivated by the most studied insect on the planet.

Tasmania an island state lying in the middle of the Roaring Forties at latitude 42° South is blessed with widely dispersed and large areas of undisturbed native flora, some unique to our island home. With strong prevailing winds and often short seasons Tasmanian beekeepers have adapted their bee husbandry to optimise available honey flows. Our native flora combined with garden flora, very useful environmental and agricultural weeds as well as horticultural and agricultural crops all provide a good floral basis for honeybees to thrive.

*Leatherwood, Eucryphia lucida a medium size tree in our temperate rainforest and wet forest provides the backbone of the industry and produces the iconic leatherwood honey.*

The Tasmanian apiary industry would not have grown to its current size without a very unique rainforest tree, Leatherwood Eucryphia lucida. It flowers most years and continuously for weeks to months throughout the summer depending on the weather conditions. This rainforest tree inhabits large areas of south west, west and northwest Tasmania and produces the iconic leatherwood honey, the mainstay of the apiary industry.

European honeybees were first introduced to Tasmania in 1834 and the Italian race Apis mellifera ligustica  in 1889, now the favoured bee in the industry.

The apiary industry has grown significantly since the early introduction of bees, originally suffering a summer lack of flowering until the discovery in the early 1900's of the rainforest tree leatherwood, an often reliable source of summer honey flow.

*Bees are vitally important to all of us, many horticultural crops are reliant on managed and feral honey bee pollination, with approximately 65% of agricultural crops responding to honeybee pollination. We need healthy bees!*

The apiary industry in Tasmania consistently produces above its weight, approximately 2% of the nation's beekeepers producing 4% of the national honey crop. Tasmanian beekeepers tend to consistently receive higher prices for their honey, most likely due to the demand for our unique leatherwood and the increase in production of Manuka, Leptospermum scoparium, a Tasmanian native.

As the health benefits of honey become more known, the demand is outstripping supply.  Notably the non-peroxide antimicrobial activity of some honey, especially Manuka, and an increase in research discovering high peroxide activity and antioxidant levels in Leatherwood and other honey.

Globally there is an increased awareness of the significant part honeybees play in providing our food and the plight of the honeybee.  Likewise the interest in beekeeping as a hobby has hit new heights.  Beekeeping courses are booked out, even here at the end of the earth!

*Lifetime learning as two veterans of the industry discuss the finer points of queen breeding.*

This book is designed to help you start and continue as a good beekeeper. It covers the topics and issues that you need to understand in the context of the Tasmanian environment and other cool temperate climates. It should be considered a companion to the many great and detailed books on beekeeping that are available.

A word of caution as your embark on this fascinating journey; most people start beekeeping because of the bees and leave because of honey. James Tew (in Flottum, 2010).  The best advice is plan and prepare, you are dealing with a super-organism, the colony, it will continue to grow and increase while food sources and conditions are favourable. As with any practical skill it is better to inform yourself before commencing. Beekeepers are very generous with their knowledge, I encourage you to join your local branch of the Tasmanian Beekeepers Association, or a beekeepers group wherever you are and get to know some beekeepers. You can be assured that you have embarked on a journey of lifetime learning.

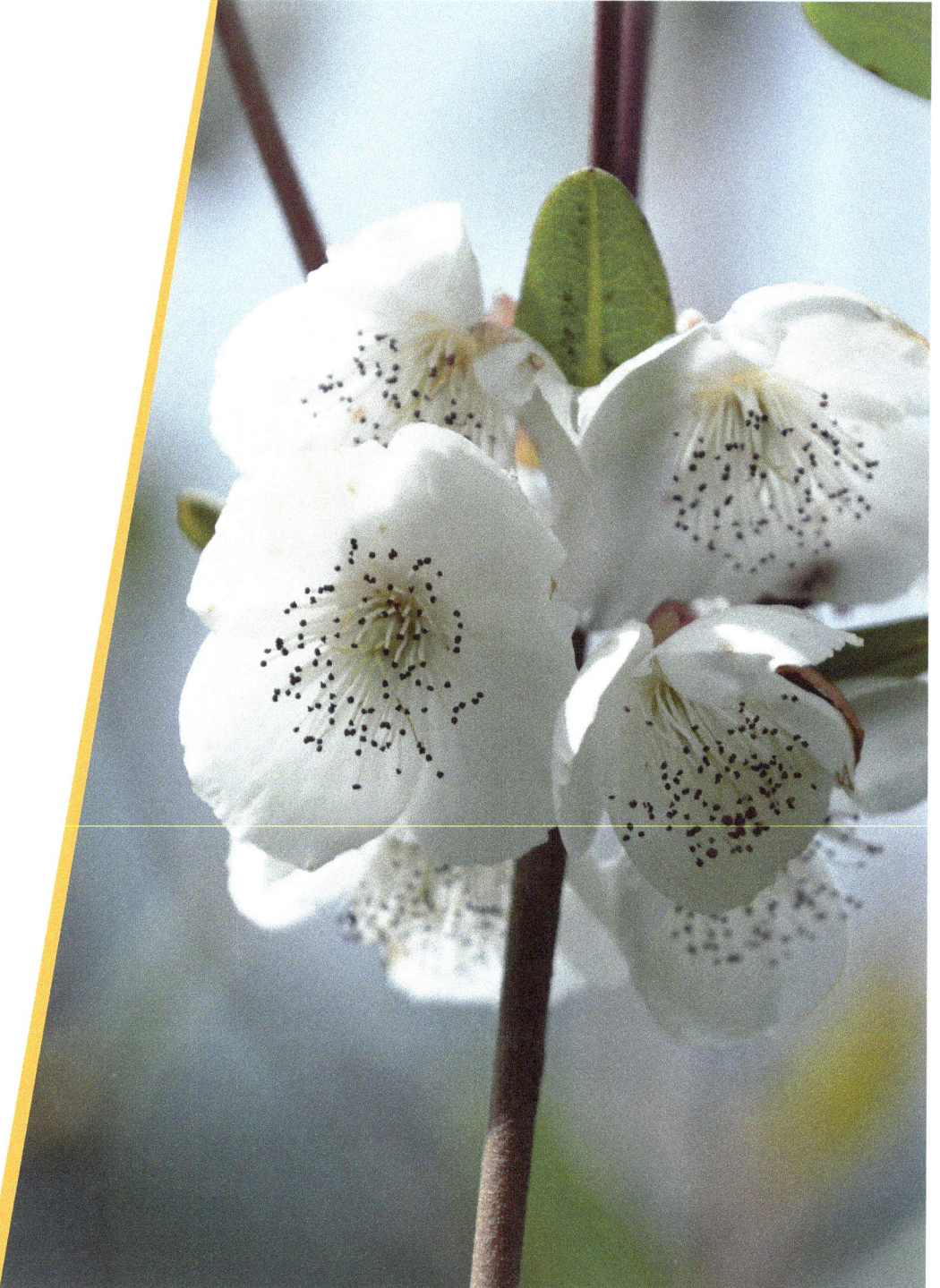

You're at a good place. Before you start buying equipment or bees it is most important to inform yourself. This book provides a Tasmanian focus with unique information on the floral resources and seasonal management issues with beekeeping principles that are applicable in most cool and temperate climates. The Bibliography lists a number of very useful books, papers and web sites that can also guide your decision making.

Join your local beekeepers association, it is full of friendly knowledgeable beekeepers, amateur and commercial who can guide you, answer your questions and remember there is no such thing as a bad question. Ask away!

Undertaking a course for beginners is also worthwhile but at some stage you need to make a start. Covering the basics before you start can help you navigate issues you may observe and questions you will have.

Key points to remember:

- Have a site selected to place your hive or hives (see Hive Location).

- Plan your visit, what do you intend to do, to see and what to take.

- Make a check list, nothing worse than arriving without your smoker, matches, hive tool, veil or gloves.

- Ensure your personal safety, that of anyone with you and your neighbours.

- Don't use strong smelling products, dark or woollen clothing and suede. Patting the dog or horse before you leave is not advisable.

- Record the date, the weather, where you are going, when you're there, what's in flower and observations of your unopened hive and then once you open the hive what you see and do.

- Always take at least 5 litres of water to wash sticky hands or gloves but especially in the summer to extinguish your smoker fuel and any fire that could start.

*A well protected location with easy access close to a source of clean water is essential.*

## CHAPTER TWO

Initial Gear

(See Equipment in chapter six for more detail)

Safety first. Your personal protective equipment (PPE) is essential and while you can make do, it is advisable to purchase quality proven clothing.

- Overalls or a jacket with an attached hood.

- Leather gauntlet style gloves.

- Gaiters: optional but provide extra ankle protection.

- Hive Tool: stainless steel specifically designed for hive management.

- Smoker: Stainless steel with a bellows often made from leather, the biggest you can afford.

Hive Parts

You will initially need boxes for your brood and their stores and some extra for a honey flow. These boxes are called supers and in Tasmania the most commonly used are Ideal supers, about half the depth of full-depth supers commonly used on mainland Australia. The advantage to using ideals is their weight for handling when full of honey, 12-13kg instead of 25-30kg for a full-depth.

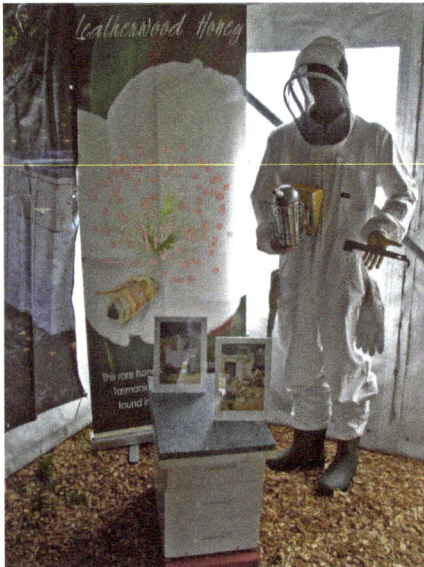

*Personal protective equipment essential for beginners: a veil and beesuit, gauntlet gloves, gaiters and sturdy footware.*

*Ideal supers are most common in Tasmania, when full of honey weighing about 12kg, a safer option.*

- Use five to six assembled and painted ideal supers and eight frames with foundation per super, a waterproof migratory or telescopic lid, a bottom board and an em-lock.

- Order a nucleus hive ; a laying queen with three frames of brood and stores.

- Also, it is better to have two hives than one as you can undertake different manipulations and see what is working the best.

*A new hive being prepared from a nucleus hive by master beekeeper Laurie Cowen. Beginners should always wear their protective clothing.*

*Each frame from the nucleus is checked for the queen and her vigour. And here she is, your new queen identified by a safe coloured marker on her thorax.*

Some commercial beekeepers use full depths for their brood and others believe that all full depths provide a much more efficient production system. These operations are usually on pallets with some form of mechanised lifting.

Now begin the journey it's the greatest way to learn.

All the best with this great adventure of learning, relationships, bees, flora and honey. Remember, one hundred different beekeepers, one hundred different ways to keep bees.

The following description applies mainly to worker bees, infertile females. The other female is the queen, a specialized female bee responsible for the reproduction of the colony. Male bees, drones main role is to mate with a virgin or newly mated queen.

Honeybees, Apis mellifera belong to the large insect family, Hymenoptera that includes other bees, wasps and similar insects. Similar to other insects, bees have three main body parts, a head, thorax and an abdomen.

The head: this includes the two compound eyes and three simple eyes, two antennae, feeding structures: the tongue (proboscis) and the jaws (mandibles), the brain and a food production gland all neatly packaged into 3mm x 2mm. Of greatest importance is the pair of strong jaws that the bees use to chew pollen and in the case of workers shape beeswax. The proboscis is like a drinking straw

## HONEY BEE
### Apis Mellifera

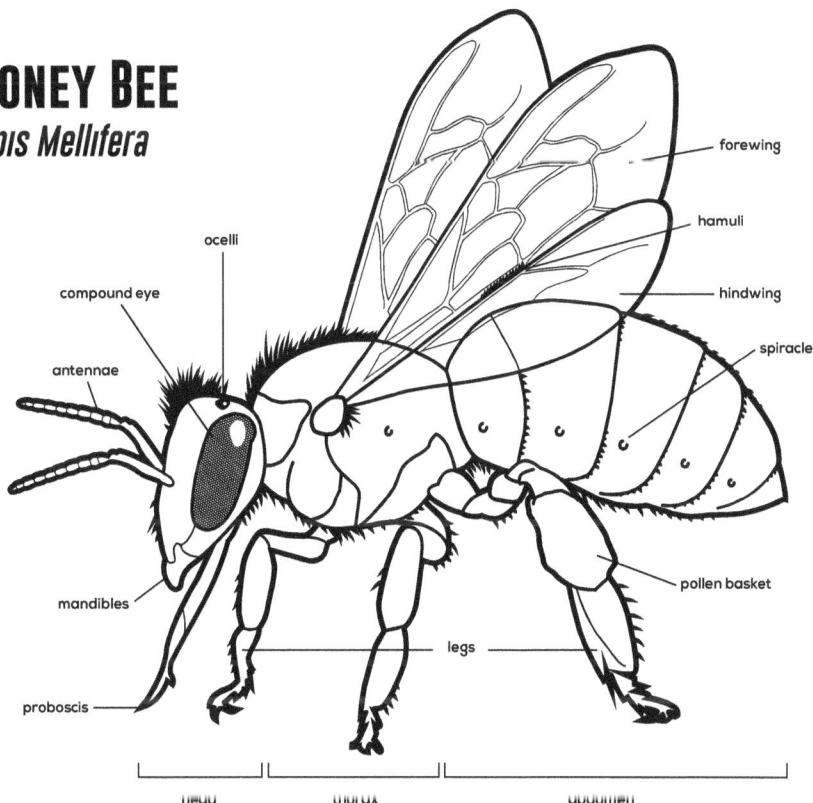

forewing

hamuli

hindwing

spiracle

ocelli

compound eye

antennae

pollen basket

mandibles

legs

proboscis

head    thorax    abdomen

*This diagram shows the main elements of a worker bees anatomy. Marc LeRoux*

attached to a pump, when in use for taking up nectar or water it is held steady by the mandibles, when not in use it folds back under the head.

The compound eyes are on either side of the head and contain between 6500 to 9000 facets depending whether the bee is a queen, drone or worker. The compound eyes perceive colour, light and directional information from the sun's rays. Bee colour vision includes, violet, blue-green yellow and orange, they can't perceive red but can see ultra-violet, invisible to humans. The simple eyes on the top of the head, ocelli are light sensitive lenses that help determine which way is up

*The compound eyes of workers and drones differ with the eyes of the drone meeting at the top of the head. Compare the size of the drone and worker, observe the general anatomy.*

The thorax: is the middle section and contains two pairs of wings and three pairs of legs that have specialized structures and hairs that assist the bee with cleaning itself, cutting wax and collecting and carrying pollen. 'The thorax is the engine room, with powerful muscles that provide motion to the legs and wings and energy for heating the colony'. The brood is kept at a constant 34°C to 35°C and if the hive drops below 8°C, the bees will die.

When the wings are folded back over the body the wing muscles can be moved to generate heat without moving the wings. The wings are moved by the muscles changing the shape of the thorax, when it is compressed the wings move upwards and when it is released they moved downwards. The wings are fine membranes stiffened with delicate veins. The rear wings are not powered and are joined to the forewings as they move across the body at the time of flight by minute hooks called hamuli, providing a larger wing area. The thorax is armour plated and perforated with three pairs of holes called spiracles, part of the bee respiratory system.

*The queen has a longer slender abdomen. David Barton Photography*

The abdomen is the longer back end of the honey bee, connected to the thorax by a waisted narrowing. The abdomen consists of six major segments and contains the honey sack, the digestive and reproductive organs, the heart, wax glands, the Nasonov gland and the sting. The sting is situated at the back of a workers bee's abdomen and is used for the defence of the colony. The sting is barbed and locks into the skin and is left behind with its venom sac and muscles if the bee flies off and will continue to penetrate and pump venom if not quickly removed.

A queen abdomen contains the ovaries for egg productions, a storage sac, the spermatheca for drone semen, no wax glands and she has a barbless sting used for defence by killing another queen. Drone abdomens contain male reproductive organs, no wax glands and do not have a sting.

The bee stomach or ventriculus is where digestion of pollen and nectar occurs. Collected nectar, honey dew and water are temporarily stored in the honey stomach, a crop like structure that sits in front of the ventriculus. The honey stomach contains beneficial bacteria Lactobacillus spp, important in the formation of bee bread and protection of honey. Pollen is filtered from the honey stomach and passes by a valve into the ventriculus allowing pure nectar to be regurgitated and passed on to young hive bees that add enzymes to it, remove some moisture and store it in cells for final drying and turning into honey.

The Nasonov gland is on the top of the abdomen near the rear and produces a pheromone mix used for communicating, attracting other bees or leaving a scent trail while flying. Alarm pheromones are released by the Koschevnikov gland near the sting shaft. Isopentyl acetate (IPA), the dominant chemical in a mix, is emitted at the time of stinging to attract other bees to the sting and cause defensive behaviour, it smells like bananas and is masked by smoke. Another alarm pheromone of different chemical structure, 2-heptanone, is emitted by the mandibular gland more commonly used at the hive entrance against smaller intruders that are bitten by the mandibles as they are too small to sting.

Sexes and Castes

It is important to understand the structure of the colony and how the different types of bees function and interact with each other, as a group and how they respond to their environment. A good place to begin is with the individual bee types in the colony.

There are two sexes in a colony, females (queens and worker bees) and males (drones). For a colony to survive it must have four components that change with the seasons;

- a queen, (the fertile female),

- workers bees, (infertile female),

- drones, (fertile males, in late spring early summer, none in winter),

- and a brood nest with eggs, larvae and pupae in the brood comb (none in winter).

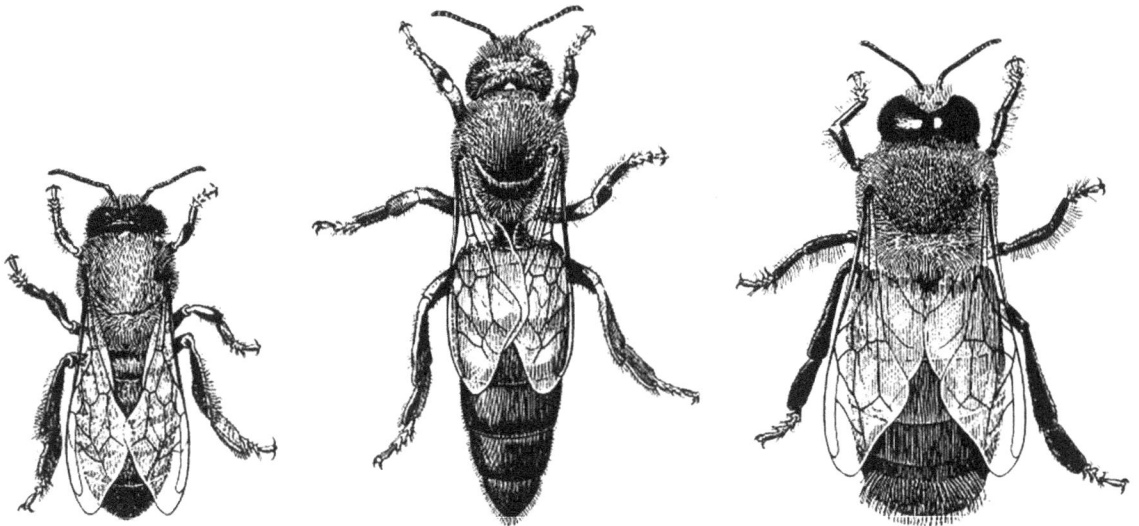

Worker          Queen          Drone

*Illustration taken from Beekeeping in the United States USDA Ag Handbook 335 .*

*Every hive or colony must have one egg laying queen to survive.*
*The queen is the mother of the whole colony.*

The Queen

(see Queen Rearing in chapter fourteen)

Every hive or colony must have one egg laying queen to survive. The queen is the mother of the whole colony. As illustrated the queen is the longest of the bees with a long slender wasp like abdomen, but with a head and thorax a similar size to a worker bee.

Any larva that hatches from a fertilised egg up to three days old is potentially a queen. A larva becomes a queen because of the treatment given to it by worker bees and the bigger cell it grows in. This ability to raise a new queen is a survival mechanism for the colony. A queen can die, be lost, injured or killed by a beekeeper, is laying poorly or is too old. "Finding the queen is the graduation of the beginning beekeeper, it's a skill you must acquire". Once found and handled without damage, she can be marked on her thorax for ease of future location.

The queen is normally the only fertile female in the colony and under ideal conditions can live up to five years. Modern practice to maintain hive vigour and prevent swarming has the queen replaced with a younger fertilised queen on a regular basis, often between one and two years.

The life of the queen begins as with other female eggs, however she has been laid in a larger queen cell when the colony is swarming or superceding. An emergency

*A marked new queen in a nucleus hive ready to build a new hive or strengthen a struggling one.*

*A comparison of worker, drone and queen brood cells. Workers are smaller, more numerous and "flat capped" drones are larger and domed and the open queen cells can be the size of a peanut.*

*Queen cells typically constructed at the bottom of the frame with the entrance facing downwards.*

queen cell starts as a worker cell with the larva floated on royal jelly and the cell progressively extended. On the third day of incubation the miniscule eggs hatch and the larvae are continuously fed on royal jelly by worker house bees making up to a thousand visits to feed them. Royal jelly is an enriched nutritious mix of protein rich pollen and carbohydrate laden honey and includes special enzymes produced by the house bees. The diet changes on the last two days of larval life with royal jelly and the addition of honey and high levels of juvenile hormone (Sammataro & Avitabile 2012). This high quality royal diet ensures complete development of the reproductive organs and the pheromone and hormone producing glands of the new queen.

Queen cells are larger than worker bee cells providing room for her full development. Typically the size of a peanut or acorn shell they hang down and are built at the bottom of a frame on the comb or in the space between adjacent comb with the open end pointing down.

*Worker bees, infertile females, make up the majority of bees in the colony and do the work of cleaning, rearing, guarding, tending the queen, making wax, building comb, filling cracks and holes with propolis, foraging for pollen, nectar, water and propolis and homeostasis (temperature control) of the hive.*

### Worker Bees

Worker bees, infertile females, make up the majority of bees in the colony and do the work of cleaning, rearing, guarding, tending the queen, making wax, building comb, filling cracks and holes with propolis, foraging for pollen, nectar, water and propolis and homeostasis (temperature control) of the hive. They begin as a fertilised egg that also incubates for about three days before hatching to a laval stage, fed on royal jelly for two days then their diet changes completely to the clear liquid produced by the hypopharyngeal glands and again after a few days to honey and pollen. On the fifth day the cell is capped and the larva turns to a prepupa before spinning its silk cover, pupating when it develops into a bee, finally emerging on day 21.

Hygienic traits are inherited so it is important to look for hygienic strains of bees. Hygienic bees tend to prevent the colony from succumbing to many different infections and disease as they quickly remove dead and dying bees and sources of infection. They remove debris and clean the inside of cells ready for egg laying, pollen and honey storage.

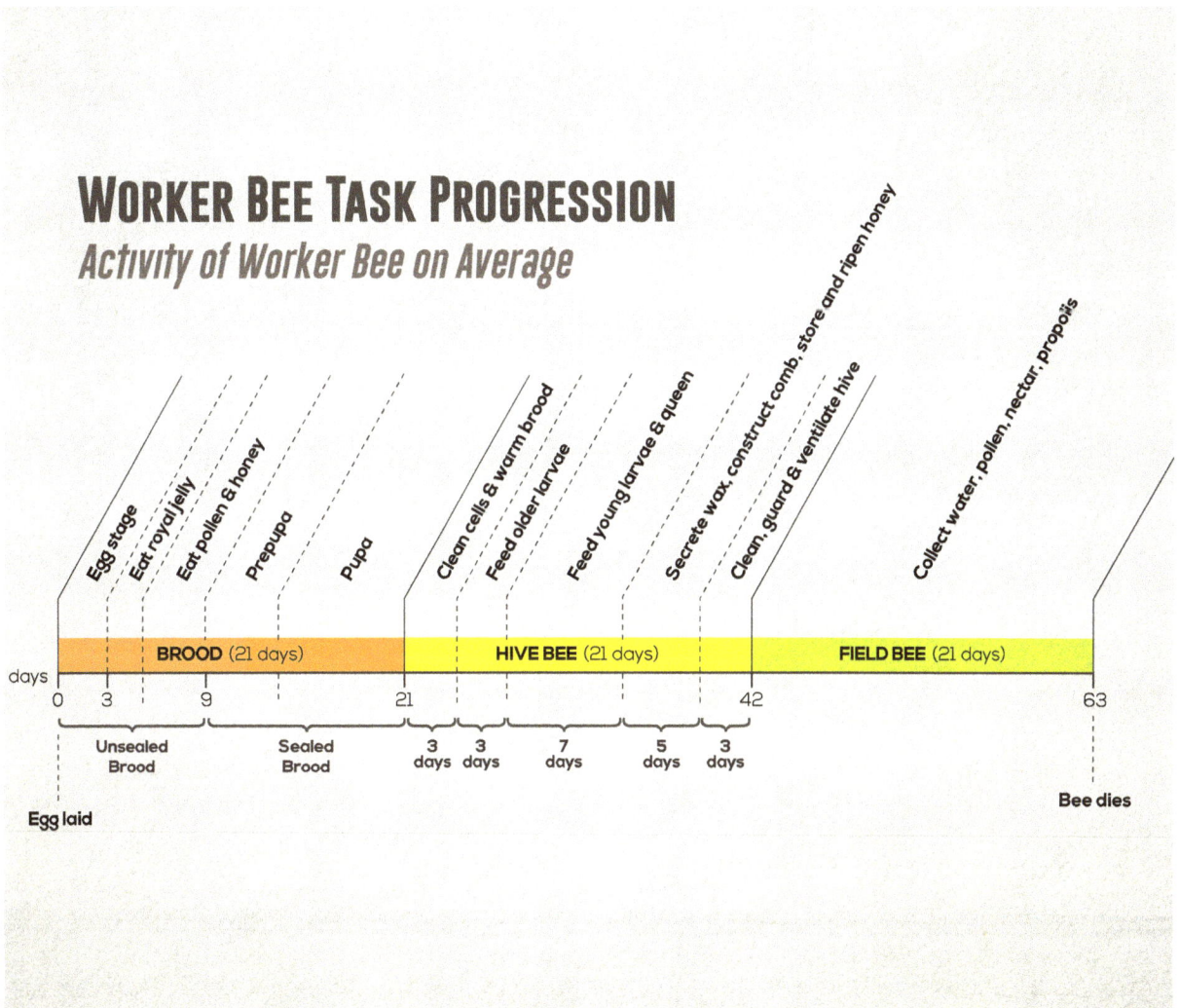

# WORKER BEE TASK PROGRESSION
## Activity of Worker Bee on Average

Egg stage
Eat royal jelly
Eat pollen & honey
Prepupa
Pupa
Clean cells & warm brood
Feed older larvae
Feed young larvae & queen
Secrete wax, construct comb, store and ripen honey
Clean, guard & ventilate hive
Collect water, pollen, nectar, propolis

| | BROOD (21 days) | | HIVE BEE (21 days) | | FIELD BEE (21 days) |
|---|---|---|---|---|---|

days
0   3       9           21      42              63

Unsealed Brood   Sealed Brood   3 days  3 days  7 days  5 days  3 days

Egg laid

Bee dies

*Marc LeRoux*

# CHAPTER THREE

*Workers going about their duties. Cleaning cells and feeding larvae.*

The tasks of a worker bee changes as different glands develop. The first ones to develop after emerging are the hypopharyngeal glands in the bees head that produce brood food for larvae and food for the queen. About 14 days after emergence the wax glands have developed and the role changes to secreting wax and constructing and repairing comb. Enzymes secreted to produce honey from nectar are from the hypopharyngeal gland as it further develops. Poison or venom glands are the last to develop when bees begin protective duties guarding the hive. Finally once the bee is fully developed and able to defend itself they become foragers, leaving the hive to collect, nectar, pollen, propolis and water.

Scout bees 'spy out the land' looking for rewarding floral sources based on flower colour, shape, detail and aroma. Once a scout has collected from a new source it circles the floral source to aid in location and flies back to the hive. The ability to memorize different terrain and landmarks is crucial for foraging success and colony survival. The quality of the nectar collected by scout bees is tested by receiving bees, if it is an acceptable quality and there is enough room in the hive the location of these nectar and pollen sources is communicated on the 'dance floor' an area near the entrance of the hive.

Location of new forage sources is undertaken by several 'dance' movements that relate to location, a circular dance for food sources within 100 metres and the 'waggle' or figure of eight for more distant food. The dancers also vibrate, and that sends information to other bees through the vibration of comb and pass on the odour of the flowers, as well as providing minute nectar samples for attendees. In this way the scouts and foragers recruit more foragers. This communication process was first discovered by the Austrian researcher Dr. Karl von Frisch in the late '60's for which he was awarded the Nobel Prize.

The diagram helps interpret the 'waggle dance'.

*Figure-Eight-Shaped Waggle Dance of the Honeybee (Apis mellifera). A waggle run oriented 45° to the right of 'up' on the vertical comb (A) indicates a food source 45° to the right of the direction of the sun outside the hive (B). The abdomen of the dancer appears blurred because of the rapid motion from side to side.*
*CC 2.5 Design by J Tautz and M. Kleinhenz, Beegroup Würzburg, Chittka 2004*

Drones

The drones main function is to mate with a virgin or newly mated queen. They also assist with thermo-regulation of the hive and their presence is thought to provide a cohesive element to the colony. Drones make up to 15% of the colony population and can be a considerable drain on resources.

*Drone brood, prominently domed, clearly larger than worker brood. Bigger cells allow them to grow*

> *Drones are the largest bees in the colony, beginner beekeepers often mistake them for queens.*

Drones are the largest bees in the colony, beginner beekeepers often mistake them for queens. The main obvious difference is the end of the queens abdomen is pointy, whereas the drones is blunt-rounded, with a heavier body and the drones head is bigger than both the queens and worker bees. The drones' compound eyes meet at the top of its head, it has no stinger, no pollen baskets or wax glands and never collects food or feeds young bees.

Drone larvae hatch from unfertilised eggs laid in larger drone cells either by a mated queen, a failing queen or in her absence an infertile laying worker. After the fourth day drone larvae are fed a modified diet of an increased quantity of honey and pollen. Drone cells are capped after six and a half days of feeding and the

*Drone pupae exposed during first spring hive inspection.*

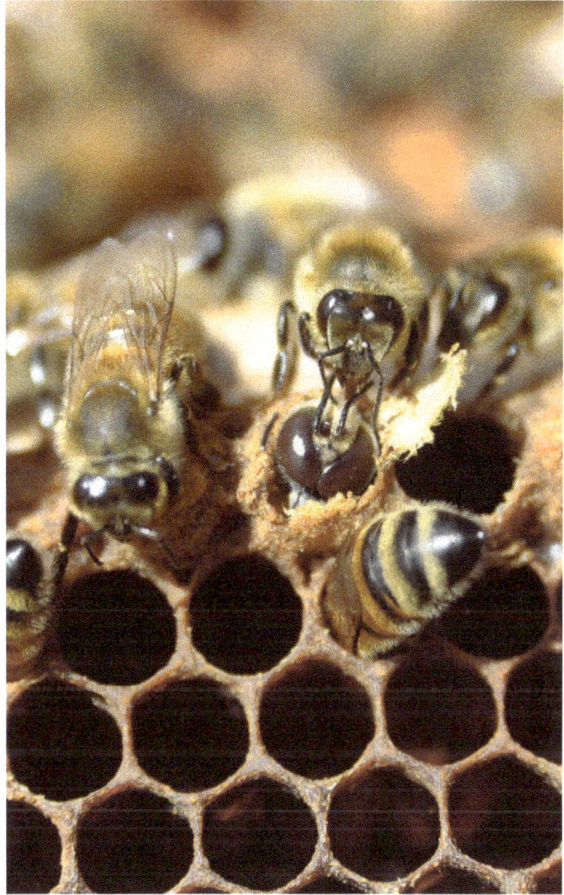

*A drone emerging being helped by worker bees.*

wax caps are prominently domed, unlike worker cappings that are only slightly convex. Once hatched, drones stay in the hive for six days initially being fed by worker bees, then begging for food. Then strong enough to take some 'training' flights, eventually able to fly to the Drone Congregating Areas (DCA's) where they congregate in flights with other drones in waiting for virgin queens. DCA's are typically ten metres above the ground where there is a slight updraft, possibly assisting their flying.

Only the strongest drones get to mate with a queen, a once in a lifetime event that leads to death, once the sperm is delivered, the drone is instantly paralysed, falls to the ground and dies. The copulation is done in flight and the queen will mate with at least ten drones over several days until her spermatheca is full. This provides the sperm for fertilising the eggs she will lay over her lifetime.

# LIFE CYCLE OF THE HONEY BEE
## Apis Mellifera

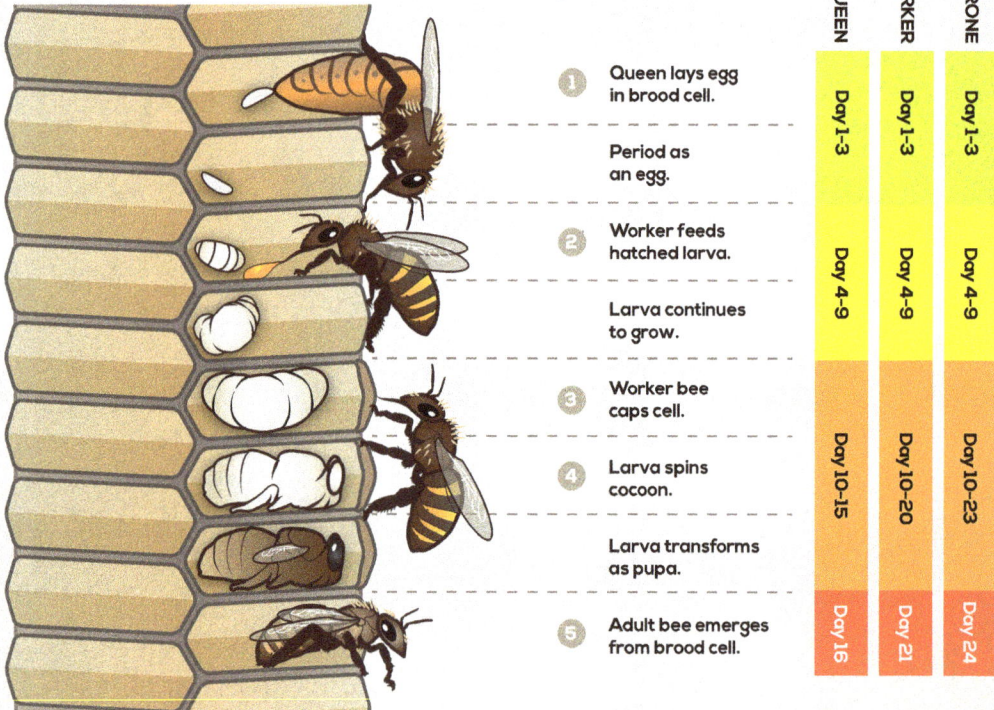

| | QUEEN | WORKER | DRONE |
|---|---|---|---|
| 1 — Queen lays egg in brood cell. | Day 1-3 | Day 1-3 | Day 1-3 |
| Period as an egg. | | | |
| 2 — Worker feeds hatched larva. | Day 4-9 | Day 4-9 | Day 4-9 |
| Larva continues to grow. | | | |
| 3 — Worker bee caps cell. | Day 10-15 | Day 10-20 | Day 10-23 |
| 4 — Larva spins cocoon. | | | |
| Larva transforms as pupa. | | | |
| 5 — Adult bee emerges from brood cell. | Day 16 | Day 21 | Day 24 |

*Average development time of a European Honey Bee. Marc leRoux*

*Eggs appear as tiny white elongated objects lying upright at the base of the cell. David Barton Photography*

*Uncapped larvae.*

*Larger drone larvae compared to the much smaller worker larvae.*

## Life Cycle

The diagram provides a summary of the development time of the different stages by caste in the honeybee life cycle from egg to emerging adult. The images give some indication of the difference between cells, the acorn shaped downward pointing queen cell and the slightly larger dome capped drone cell. The first three days are essentially the same however worker bees may have pollen in their diet within the three day period.

## Honey Bee Races

There are many varieties of honey bees but the most commonly used bees in Tasmania are the Italians, A. mellifera ligustica. The first bees introduced 'English Black', A. mellifera mellifera, also called German Black or European Dark bees form the basic genetics of the feral or wild European honeybee populations. While perhaps better adapted to the Tasmanian climate they are more difficult to manage, are more aggressive and produce less honey.

The table provides a comparison of three subspecies used in Tasmania and Russian bees that have been used in US breeding programs for varroa resistance.

# CHAPTER THREE

| Trait | Italians<br>*Apis mellifera lugistic* | English/German<br>Black Bees<br>*Apis mellifera mellifera* | Carniolans<br>*Apis mellifera carnica* | Russian<br>*Apis mellifera* |
|---|---|---|---|---|
| Origin | Southern Italian Alps | Britain, Germany, middle Europe | Slovenia, Southern Austrian Alps, Northern Balkans | Primorsky Coast |
| Colour (ave) | Light 4-5 bands on abdomen, Light brown-yellow | Very dark | Grey to dark | Grey |
| Behaviour/Disease Resistance | Can show excellent cleanliness (a trait to breed for) | Better hygiene but susceptible to brood diseases | Resistant to some brood disease | Very hygienic, show some resistance to varroa. Introduced into USDA breeding program in 1970's |
| Gentleness/Ease of handling | Very good | Aggressive | Extremely gentle can be worked with very little smoke | Tend to be aggressive. |
| Spring build | Tend to build early, late autumn build consumes stores, rapid spring build very useful for early flows and pollination but can cause food shortage | Poor, slow spring build | Very good | Average, a slow start but can have explosive growth leading to swarming if not well managed |
| Overwintering ability | Larger populations, no tight winter cluster uses more stores to create energy | Conserve winter stores | Smaller populations use less stores | Very good at conserving stores with a very small winter population |
| Swarming potential | Lower, manage laying space | Low tendency | Prone to swarm | Explosive growth can lead to lack of room and swarming |
| Honey Production | Large surpluses, excellent foragers, excellent producers | Average honey producers | Above average but not as productive as Italians, swarming causes poor honey yield | Average honey production |
| Propolis (excess propolis use makes management difficult – sticky/glued) | Litlle propolis | Average propolis use | Low propolis use | Average propolis use |
| | Propolis may be the first line of defense against disease/mites and breeding for lower production may have a negative affect ( from Spivak) | | | |
| Other | Globally most commonly used commercial bee. Very adaptable. Queen easy to find as different colour. Tend to drift and rob (passing on disease) | Fly in poor conditions, Longer lived. Difficult to manage as very agitated on comb. Tend to become honey bound | Queens difficult to find. Good for colder climates. Very pollen dependent. Less drifting & robbing. White wax capping due to airspace | Brood affected by flow. Queens cells always present |

# CHAPTER FOUR

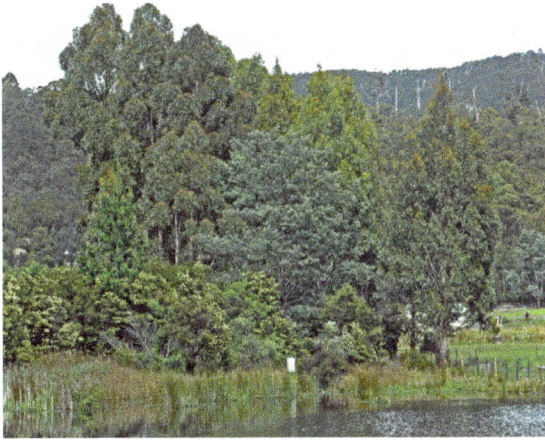

*A well sheltered site close to a source of clean water with abundant flora.*

Finding an appropriate location for your hive is important before you have your colony of bees. There are some fundamental considerations that apply in case you have your bees in your backyard, some other urban environment or in a more rural setting. Most requirements of hive location are equally important, but WASP( water, accessibility, safety and protection) sums up the physical attributes of a location. What plants are flowering or will flower is also important and is covered in the points below;

- Water: bees need water to regulate hive temperature (spread out and cooled by rapid wing fanning, the bees air conditioner), raise brood (used by nurse bees to produce royal jelly made from pollen, honey and water in the hypopharynx), dilute honey and liquefy crystallized honey (stored honey in winter form from many floral sources crystallizes, and needs to be reliquefied to feed bees and brood). A strong colony will need at least 1litre of water per day. Ensure you provide water close to the hive as bees will fix in the nearest available water. You don't want them in the neighbours pool.

- Accessibility: the hive needs to be readily accessible by vehicle and flat or slightly uphill so you carry the heavy honey supers downhill.

- Safety: ensure your bees are not going to create a nuisance to people or animals.

- Protection: It needs to be protected from the cold prevailing winds, often northwesterly in the equinox (September-November), shading from late afternoon sun, but not all day. Natural or available protection such as your fence, house, shed, a hill, windbreak or forest edge all provide some measure of shelter. Screening from neighbours and vandals can save you a lot of problems.

- Flora: urban flora is usually not an issue as there is something flowering just around the corner. However the modern design vernacular of austere 'spikey' plant gardens is not bee friendly. If it's a rural location you need to be aware of what is flowering and when, as you can have a big spring build, then a large population and no flowering means no food, a problem for the bees and you.

*Access is very important. An all-weather track next to your site minimises handling for small beekeepers.*

Proximity to your home location is important. Hives that are too distant tend to become neglected especially with rising fuel costs. This is also important if you want to consider "food miles" and keep it as local as possible, reducing energy consumption.

This of course may not be so important when you're chasing leatherwood or some other interesting unifloral honey flow, the sites may be a one to four hour journey. Sounds like the western world version of honey hunters.

Urban

Urban beekeeping is one of the fastest growing hobbies with courses sold out from Hobart to London and beekeeping clubs and associations growing beyond their expectations. This interest and growth can be in part attributed to the considerable media attention and Hollywood productions that have promoted the plight of the honeybee.

*A well organised apiary in a blue gum woodland, sheltered, accessible and close to a major nectar source.*

An important reference for anyone wanting to establish hives at home is the Tasmanian Beekeepers Association Code of Practice for Urban Beekeeping, It is a response to the growing popularity of urban beekeeping common to all of Tasmania and Australia. It is intended that this becomes a reference that informs a uniform approach by municipalities to urban beekeeping.

Contact your local municipal offices to find out about keeping bees in your backyard, as current regulations vary widely across the state.

Be responsible and avoid having your bees in a place where they may cause a nuisance. (Somerville 2013)

Most people who are not beekeepers or involved with bees will be concerned about

*Most people who are not beekeepers or involved with bees will be concerned about hives next door unless you explain their management and the precautions you are taking.*

hives next door unless you explain their management and the precautions you are taking. Hives should not be closer than 3m to a boundary unless it is a solid or impenetrable barrier at least 2m high.

Establish a constant water source close to your hive from the time the hive is established, preferably as capillary water in wet sand or gravel. If the bees start collecting water from a place that is a nuisance or dangerous such as the neighbours' pool or yours, they will most likely stay with that source and become a problem.

Locate your hives in a quiet part of the garden that is as sheltered from prevailing weather as possible. Face the entrance across your property, not the neighbours, a clothesline, parked cars, footpaths or roads and ideally create an impenetrable barrier at least 2m high in front of the hive entrance forcing the bees to fly up and away above head height. This is also how they will return to the hive and will be unlikely to cause a nuisance.

Screening your hives will help reduce neighbour concerns and maintain a higher flight path. Hives should not be located close to schools, childcare centres, hospitals, sporting grounds or other public facilities. (Somerville 2013)

Rural

If you are unable to keep hives in your backyard, have more hives than allowed in town or are after a particular honey flow, then you will need to seek permission of a landowner to place your hives on their land. Hobby farms are usually closer to the city boundary and their orchards and gardens benefit greatly from having some resident hives. This may also provide a good winter home for your hives if you are a honey 'hunter'. On larger properties, you can usually come to some arrangement with the landowner with a honey payment. You will need to ensure that your hives are closed with an emlock and protected from stock as cows and sheep love to rub against hives and their size will knock your hive over.

*A well located apiary close to a prickly box dominated forest. Often the source of large honey flows.*

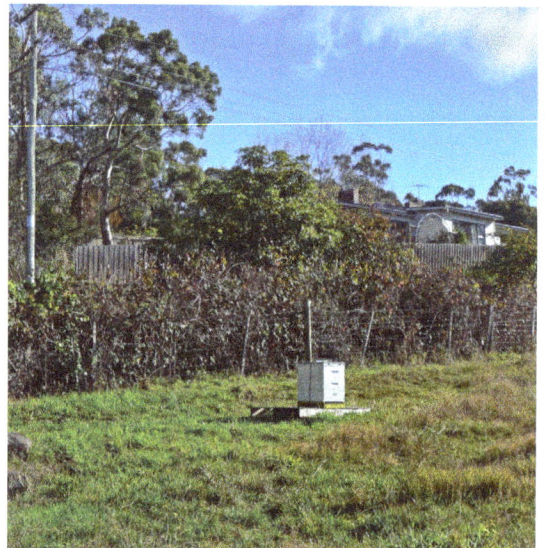

*A sheltered, sunny site on a hobby farm, away from the boundary, neighbours and home.*

If the area is known for a particular honey flow such as clover, prickly box or blue gum you need to know how many other beekeepers are accessing the area. It can become overcrowded, then no one produces a good honey crop. You can also encourage the landowner to plant bee friendly plants that can provide benefits for pollination or generally increase bee and pollinator populations and health on the land.

Stationary beekeeper

*Consider planting a floral sequence to provide you with early flowering and continuing honey flows*

Consider planting a floral sequence to provide you with early flowering and continuing honey flows. Many small urban apiaries are stationary, but larger scale commercial operations rarely are. This is possible and has been considered in both Bee Friendly (Leech 2013) and Better Bee Keeping (Flottum 2011). You may be able to lease some land, or if you own land consider planting lane ways, shelter and even a purpose planted block. This is a significant subject and if you are interested it is advisable to read what is available, then talk to beekeepers in your area. You will most likely have to plant a mix of local natives, other Australian natives and exotics. You can get plenty of clues for exotic plantings from what is flowering and attracting bees in your location of interest or an area that has a similar climate.

*A Leatherwood site, close to a main road, and sheltered from the wind, but possibly too much shade.*

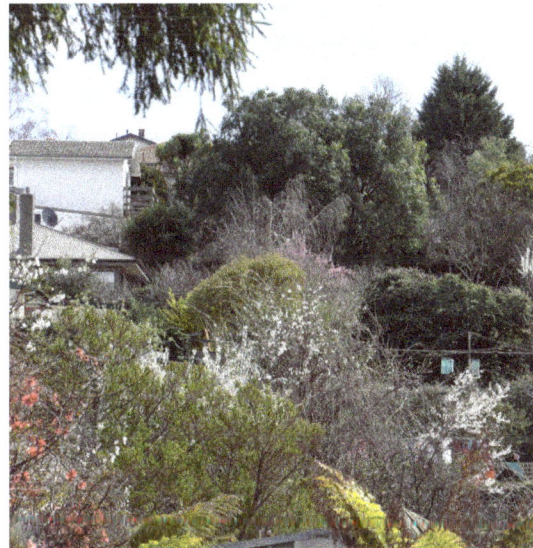

*An example of the diversity of flora. Spring time in Launceston, northern Tasmania.*

Rooftop Hives

There is a growing trend in cities around the world to have hives on roofs, generally away from people, but with their own set of issues. An unwritten competition seems to exist to see who can get their hives on the most iconic buildings, from famous museums to other historic sites. Even some hotels such as the Fairmont Group have established apiaries on their roofs to provide in-house honey.

Melbourne City Rooftop Honey has provided a great model and have a vision to bring bees back to the cities and suburbs of Melbourne. The aim is to increase interest and awareness and reduce the distance from "paddock to plate". With hives on the roofs of cafes, restaurants and hotels, food miles have been reduced to metres.

While all exciting, it is difficult to move hives up stairs and in and out of windows. Contact your local beekeepers branch if you are contemplating this, there will be a beekeeper with some experience and knowledge. We don't want your lounge room full of bees or a hive to fall on the street.

*Federation Square rooftop Rooftop Honey (Federation Square rooftop in Melbournes' CBD provides a great platform for a 10 hive apiary). Melbourne City Rooftop Honey.*

*Managing rooftop hives requires attention to planning and safety. Melbourne City Rooftop Honey.*

Increased awareness of the plight of honeybees has stimulated a growing movement in urban beekeeping. Cities all over the Western world are embracing beekeeping as an interesting, rewarding hobby and one that can offer commercial benefits. Australian cities are part of the trend with bee courses sold out and hobby, urban and commercial apiaries increasing in number (Leech:2013).

Keeping bees at or near your home is a long tradition of beekeeping all around the world. Urban environments are becoming known as safe places for bees, often safer than agriculture environments, lacking widespread industrial scale spraying programs.

*Lavender provides a good source of nectar*

*An example of a great bee garden with larger clumps of individual species*

*Cities and towns provide year round diverse flora, from productive home orchards and vegetable gardens to ornamental plantings of a broad range of native and introduced plants*

Cities and towns provide year round diverse flora, from productive home orchards and vegetable gardens to ornamental plantings of a broad range of native and introduced plants. Many plants already in gardens provide excellent forage for bees, from ground covers to tall trees. We can all make a difference when making new planting choices by considering plants that are beneficial to bees. Orchards, herb and vegetable gardens are all good as most species provide nutritious pollen and nectar.

# CHAPTER FIVE

*Plums provide early pollen and nectar and the bees pollinate them giving you choice fruit*

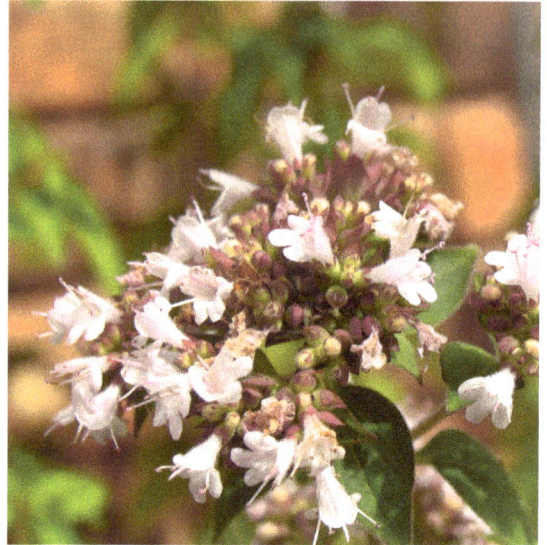

*Even herbs in a pot provide beneficial forage*

It is possible to provide a continuous flowering sequence in your own domestic garden. While bees prefer to stay at home in winter, they will venture out when conditions allow, having beneficial forage available helps the hive and conserves stores. Urban environments are often sheltered providing more flying opportunities in otherwise adverse windy or chilly conditions.

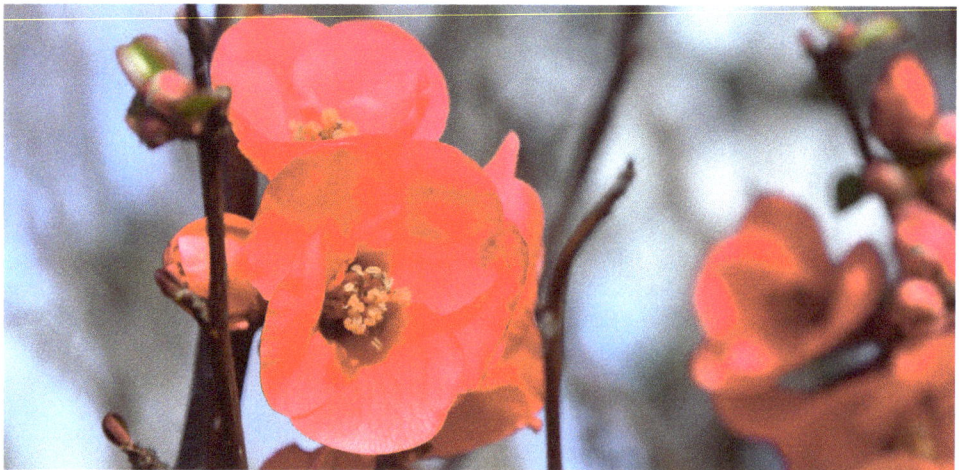

*Chaenomeles japonica is the first late winter flower providing bees a forage opportunity when the weather clears*

## Regulations

Check out the local government bylaws and regulations related to keeping bees at your home. In Tasmania this varies across the State from being fairly reasonable to prohibition. A code of practice for urban bee keeping has been developed by the Tasmanian Beekeepers Association.

There are a number of basic principles that are effectively golden rules of urban beekeeping adapted from the TBA Urban Beekeeping Code of Practice and other sources:

- Register your hive with State Growth, it is not compulsory but a biosecurity must, http://dpipwe.tas.gov.au/Documents/Beekeeper_Registration_Form.pdf

- Join your local beekeepers association or branch where you will get to network with experienced urban and commercial beekeepers.

- No hives on blocks < 400 m², see the Code of Practice for the recommended scale from two hives on 400-1000 m2.

- Try to locate the hives out of site if possible as this will minimise complaints based on fear of bees.

- Roof top hives is an increasing trend, seek advice from your local beekeepers branch.

- Place hives three metres from boundaries in a sunny sheltered spot, not near neighbours, roads or foot paths, where the bees fly across your property and not across clothes lines or cars. Bees do cleansing (poop) flights!

*A well established urban apiary, screened on all sides with a 2m fence, good access and a work bench*

- Advisable to place or grow a barrier at least 2m high in front of the hive entrance. Bees will fly back at least at this height and down to the entrance.

- Manage for swarming, regularly requeen with a docile strain.

- Provide adequate water, preferably capillary; wet sand or gravel, even a bird bath with some floating corks (Purdie:2014).

- Do not store stickies in the open or leave burr comb, propolis or other hive products around your apiary, keep it clean.

- Use escape boards to remove honey rather than shaking, brushing or blowing methods (Somerville:2013).

- Know and understand some important bee diseases; American Foul Brood and European foul Brood are notifiable diseases, if detected or suspected you need to notify the Apiary Specialist at DPIPWE.

- Smoker use is not recommended on days of total fire ban. Always have 5 litres of water available.

- Do not work your bees or rob honey when it is cold, windy or wet or when your neighbours are working or relaxing outside and ensure pets are kept inside.

- Mow when the bees are inactive otherwise you may have to use your smoker (Somerville:2013).

It can't be emphasised enough; join a beekeeping club or association or at the very least befriend an experienced beekeeper. It is a great way to learn and gain valuable insights into mistakes that you may not have to make. There is always a wealth of information about hive management, what's flowering when, where

*It can't be emphasised enough; join a beekeeping club or association or at the very least befriend an experienced beekeeper.*

*Winter daisy provides autumn forage*

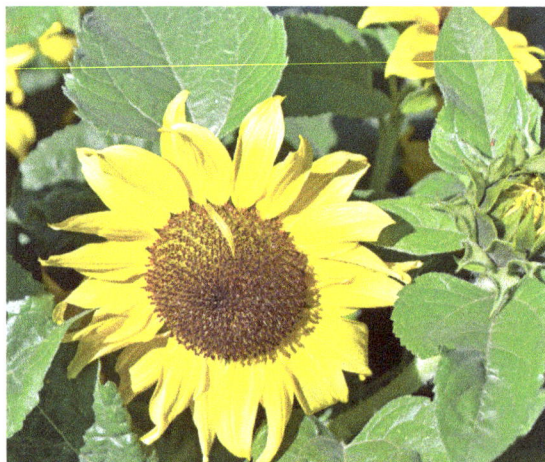

*GMO free sunflowers are a wonderful garden addition providing bee food, a joyous flower and nourishing seeds*

to buy equipment, queens and bees, how to do anything bee related. Oh yes and beekeepers are a very friendly bunch, keen to share their experience. Just remember beekeeping is a very social practice!

Almost weekly in the printed media there is an article on bees in the cities of Australia and other parts of the world.

European urban beekeepers claim it is the way of the future for them, with city environments warmer in winter, having no broad acre spraying and often year round flowering (Guest:2010 and Bethge:2008).

It's a common view that yields from urban hives are considerably less than commercial hives that follow honey flows. However, be prepared to handle a lot of honey and you may well surprise your commercial colleagues when occasionally you take over 80kg per hive from stationary hives surrounded by continuous flowering. The difference will be that following honey flows produces unifloral honey such as leatherwood and clover, while urban honey is from polyfloral sources and could be called urban honey. Your advantage is definitely less handling, low food miles and wonderful garden pollination.

Keeping bees in your urban environment is one of the few ways you can become a city farmer.

*A well structured garden with large clumps provides the best outcome for bees*

Beekeeping Equipment

With some understanding of the colony, its bees and their development it is now time to consider the equipment that you will need to protect yourself, others and to safely house and manage the bees. Basic equipment needs are few and common to all beekeepers regardless of the scale of their operation. Differences usually relate to handling larger numbers of hives, bees and hive products efficiently. The focus here is the beginner, but some idea of what larger operators use is interesting. There are a number of Tasmanian based suppliers as well as many reputable mainland suppliers used to shipping to Tasmania. Internet shopping can be attractive from a price consideration but you are encouraged to shop within Australia and Tasmania wherever possible. If you don't use it you'll lose it!

Personal, protective equipment:

It is very important to ensure you are fully protected before engaging with your hive. Clothing should be light coloured and clean. You will see some professional beekeepers without gloves and even working without veils, but they have been doing it for a long time, know what they're doing and how the bees are reacting in any given situation. To begin, wear it all!

*This basic equipment will last you years. Buy the biggest smoker you can and remember to clean your equipment.*

*Beginner beekeepers should wear all appropriate personal protective clothing.*

Bee suit with hooded veil, overalls, jacket or simply a veil. The ankles and wrists of the overalls and wrist of jackets should have elastic to help prevent bee entry. An elastic strap sewn on the trouser leg cuff of your bee suit is used to stretch under your footwear and prevents the trouser leg lifting and exposing your ankles.

- Protective gloves usually a softer leather, preferably gauntlet type with elastic tops.

- Gaiters to protect your ankles or gaffa tape strapped around the bottom of your trouser legs works.

## Handling equipment

- A smoker: Ensure a good quality one that provides enough volume for fuel and 'puff'. These are usually made of stainless steel and the bellows either a plastic product or leather. It is important to ensure they have some appropriate form of heat guard as they can get very hot.

- Hive tool: Stainless steel tool designed often with the handle end painted a bright colour for easy location after you put it down. It is used for wedging between supers to pry them apart, lifting frames and scraping wax and propolis from hive equipment.

- Bee Brush: A fine long haired soft bristle brush used to remove bees from comb.

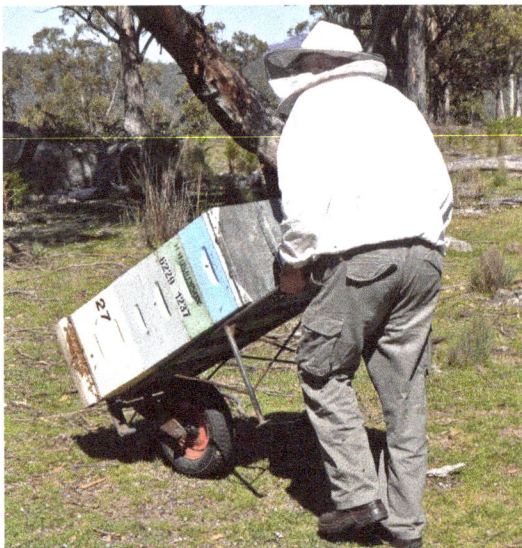

*A single wheel hive trolley handles very well on rough terrain and can convert to a barrow platform for carting loose supers.*

*Tailgate loaders are a relatively inexpensive back saving option for smaller beekeepers.*

- Hive barrow or a hand cart: A specialised single wheel hive barrow will have to be custom built, but they are great all-terrain vehicles. Conventional two wheel handcarts need larger wheels and a bigger platform to be more useful on rougher ground.

*An insdustrial crane-pallett loader makes moving large numbers of hives much more efficient.*

*A Moffat offroad loader provides a versatile industrial option that can be used in beeyards and the factory site.*

Specialised mechanical loading and unloading equipment usually means  moving more than your smaller number of hives. But remember hives full of bees and stores and supers full of honey are heavy. Always bend your knees and lift with a straight back using your legs.

The Hives

The description and discussion is based on the most commonly used hives, alternative hives and recent innovations are covered in another chapter. A basic list of components is provided and a discussion of hive structure, types and construction follows.

- A lid telescopic or migratory: A weatherproof lid that fits on top of the super.

- A bottom board: A board with cleats that forms the base of the hive and provides bees with adequate room to enter or exit. Some may have modifications to further protect from cold air entering the hive or to move the bees into the middle of the hive. Screened bottom boards, with a mesh base are also being used as biosecurity measure against small hive beetle and to assist with Varrora  management.

- Supers: These are the wooden box frames that hold the frames of comb.

- Brood supers in Tasmania: These are either two full-depth supers and frames or four ideal supers and frames.

- Honey supers: Ideal supers are most commonly used due in part to flexibility of management and their lower weight when full of honey. There are some commercial operations using full depths or Manlys to increase efficiency, note they use mechanical handling.

- Frames: Traditionally made from timber, now some use plastic, hold foundation and drawn comb and provide bee-space. (the space bees need to move freely between comb is 3/8" or 9.5mm)

- Wax or wax coated plastic foundation.

*This is a very common hive configuration in Tasmania. Some beekeeper use two full-depth supers for the brood and ideals for honey.*

- Note that hive components have traditionally been constructed from wood. Good quality durable softwood is preferred such as macrocarpa, celery-top pine or hoop pine from Queensland. Radiata pine treated with copper naphthenate (a registered chemical that has a prescribed procedure for use) and painted with a good exterior acrylic paint or paraffin and beeswax dipping provides a lower risk of residue. Plastic and polystyrene hive bodies, frames and foundation are also available.

- Queen excluders, use plastic or metal? To use or not to use: In Tasmania, plastic excluders do not conduct cold into the hive and do not bend out of shape but they are not as durable. Queen excluders prevent the queen from leaving the brood box and entering honey supers. If the brood is established in honey supers, those frames cannot be extracted. Most commercial beekeepers do not use queen excluders.

*Feral bees will make their home anywhere that is dry and has enough room. This tree has been used by bees for many years.*

## Why a constructed hive?

Bees naturally construct a home in hollows in trees, caves, building cavities, safe and dry places. Natural 'hives' contain vertically hanging comb often in a parallel series that provides a nest for brood and storage for honey and pollen. Over the millennia, beekeepers have constructed bee hives made from clay to straw, wood and plastic, even cyclone proof cement hives in the Pacific Islands.

Modern hive construction is based on the discovery of the need for bee-space by the Reverend LL Langstroth in 1852. This now forms the basis of most modern commercial apiaries. A closer to nature approach is based on the top bar hives where only a starter strip of foundation is provided and the bees build vertical comb as in a natural state. This is discussed in chapter eighteen: Alternative Beekeeping and Innovation.

The modern standard hive.

The standard box or hive now in general use, commonly known as the 'Langstroth' was initially developed by the Reverend L.L. Langstroth who patented it in 1852 and used 9.5 mm spaces between the side bars of a frame and hive wall, which later become known as bee space. The Langstroth hive consists of a rectangular body with inside measurements 464mm long, 308mm wide and 241mm deep. The box (super) has rebated ends and is built of 22mm to 25mm dovetailed pine boards. However, many commercial beekeepers in Tasmania use ideal supers for their entire apiary as they only need to carry one size super. Others use two full depths supers for the brood and a few use full depths or Manley frames.

**Australian Hive Types**
All external frames have same length and width if using same thickness timber (22.2mm thick) 508mm x 406mm.

|  | Super Style | Depth (mm) |
|---|---|---|
| Tasmania: often used for the brood | Full Depth | 244 to 245 |
|  | WSP | 193 |
|  | Manley | 168 |
| Tasmania: most common size | Ideal | 147 |
|  | Half Depth | 122 |

The floor is raised 25mm or more at the front and back so that air can pass underneath and the flat lid overlaps or fits accurately over the box on all sides to make it weatherproof.

The entrance, the width of the front, is formed automatically by nailing 10mm-25mm deep strips to the floor at the back and sides.

The eight frames (most common in Tasmania), although ten and twelve are sometimes used, run from front to back of the super. This construction provides 'bee space' between comb and sides of the hive. They are spaced so that bees can construct full frame, double sided combs for rearing the brood and storing honey, and have adequate space between the faces of adjacent combs to move freely during their work.

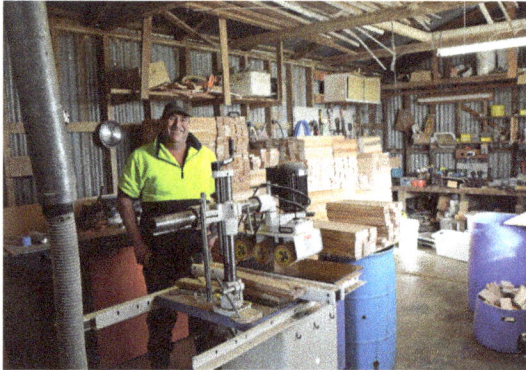

*Beekeepers need to continually repair and replace equipment and a well equipped workshop is essential*

*Ideal super construction (Ideal supers are most commonly used in Tasmania and come either with finger jointed or rebated corners).*

To ensure that the combs will be made inside the frames, a sheet of wax foundation is placed within the frame. Foundation can either be wax or plastic with bee cell bases embossed on it. Plastic foundation is coated with melted wax applied with a paint roller. This description of a hive satisfies the bare necessities of the working colony – a warm, weatherproof home with adequate space for rearing the brood

*A simple home made frame jig is a more effcicient approach than one by one. It holds 16 frames while they are stapled and assembled*

*A wiring loam holds the frame under compression while wire is threaded and secured with tacks. Relieving the compression on the end bars helps tension the wire*

and winter stores.

It is advisable to purchase your initial boxes either made up and ready to paint, or use as a flat pack ready to assemble. If time is not a constraint the latter is more cost effective and perhaps connects you more with what you are about to embark on.

*A 12 volt embedder heats the wire and melts the wax to embed it providing support to the foundation*

*A frame with foundation ready to be placed in a super and drawn into cells by the bees*

*Some beekeepers treat their raw wood with copper napthenate prior to painting*

*Painted ideal supers ready for use. To prevent them sticking to each other some insulation tape or Vaseline can be used along one edge*

Provision for the honey surplus to be robbed by the beekeeper is made during a honey flow by placing extra 'honey' supers on top of the brood chamber. Queen excluders are used by some beekeepers to prevent the queen from laying in honey supers. Other beekeepers note that using a queen excluder reduces their honey take and call them honey excluders

Hive maintenance is critical to equipment longevity, most wooden hives are now made from radiata pine which is non-durable and will decay quickly if exposed to moisture. Cracks and poor joints create more work for the bees filling them with propolis or having to increase hive temperature because cold air is coming in. Annual maintenance is important regardless of the size of your apiary.

*Annual maintenance is important regardless of the size of your apiary.*

Apart from differences in dimensions, a number of modifications and refinements are possible in hive constructions that have a place in general management;

1. Hive tops or lids. A ventilated, flush-fitting migratory lid with mats is common for local conditions though a telescopic lid made with pine or hardwood which is covered with tin is also used in many areas.

2. Completely plastic hives are entering the market. With the claims of no painting and minimal maintenance they appeal to some.

3. Frame mats or covers are essential to help maintain hive temperature, allow ventilation, prevent moisture dripping and stop the building of burr comb on the inside of hive lids. Common materials used are vinyl floor covering or table covering. Beware of carpet as bees can become stuck and will fill it with propolis.

4. Importantly the size of the covering should create a 10-15mm gap between the inside walls of the super and the edge of the mat. This applies to all super sizes and provides a cover to the frames when they are pushed together, while maintaining hive ventilation.

5. Reducing the size of the hive entrance in winter reduces cold air entry and can help prevent robbing.

6. On low-lying or overgrown sites, hive stands elevate the hive and help keep it dry. Some beekeepers like landing boards to help returning bees enter the hive while others point out that there are none in nature. Lack of landing boards

*Plastic components while not traditional provide less maintenance.*

*A simple mouse guard can be made by cutting slots in a treated pine board. This reduces any robbing impact and keeps mice out.*

makes hive loading easier. Mowing or slashing to keep the grass and weeds down helps the bees locate and re-enter the hive if they fall short. It is also a fire protection measure. Second hand timber or plastic pallets are readily available and provide an excellent working platform for small apiaries.

7.  Strapping the hive with a metal emlock or a webbing strap secures the hive components. This is essential when moving and transporting the hives.

8.  Mouse guards: An entrance cover with a series of 10mm slots or holes to prevent mice invasion.

The Smoker

Smokers come in a variety of sizes and all perform the same function, containing smouldering fuel producing cool smoke to manage bees. They are usually made of stainless steel, have a bellows made of leather or vinyl, a lid with nozzle, a fuel grate and some form of heat guarding. Its best to buy the largest size available as the fuel lasts longer and they are easier to light.

*The best fuel to use is the one that works best for you*

Fuel

The best fuel to use is the one that works best for you. Some commonly used fuels include;

- Dry pine needles, these tend to produce a resinous build up.

- Dry sheoak needles or leaves, these burn slowly and don't produce any resin.

- Straw, clean burning with no resin.

- Sugar cane mulch, clean burning with no resin.

- A mix of wood shavings and saw dust.

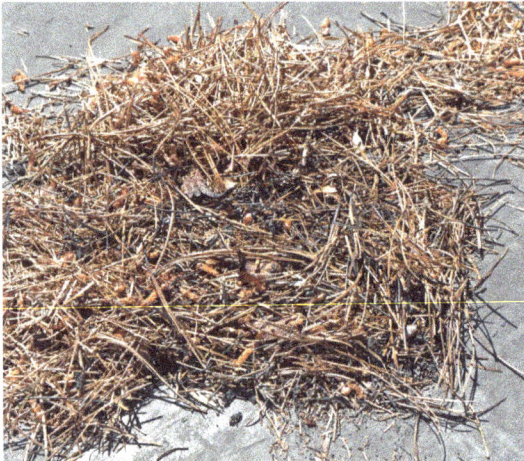

*Dry pine needles provide a readily available good fuel.*

*Sugar can mulch leaves no residue and is a long burning fuel*

There are many others so find out what is readily available, try a new fuel several times before using your smoker for the first time.

Lighting Your Smoker

- Clean your smoker nozzle of any resin build-up.

- Light a small piece of 'scrunched' up newspaper and drop it into your open smoker.

- Add a small amount of your fuel while giving the smoker a few puffs.

- Continue adding fuel while puffing your smoker, but don't pack it too tight, you want air to move through the fuel to continue the burning process.

- When a lot of smoke continues to rise after puffing and it keeps burning for a minute or so, close the lid.

- A smoker when properly lit should smoulder on its own for quite some time. Give it some rapid vigorous puffs when you pick it up, but don't overheat it. The aim is to produce continuous cool smoke to gently manage your bees.

## Using your smoker

Cool smoke blown first into the entrance masks the alarm pheromone emitted by guard bees, some light puffs over the top of the frames then also stimulates the bees to gorge on honey as if they are being confronted by a wild fire. Bees will also move off the top bars and into the hives as gorged bees are more docile. Oversmoking or hot smoke can have the opposite affect and upset the bees. Be gentle in your approach, light puffs, only more if needed.

If you get stung, after scraping the sting off, a few puffs of smoke on the sting site helps to mask the alarm pheromone emitted by the stinging bee, the pheromone is a chemical attractant to other stinging bees.

## Fire prevention

Summers in Tasmania can be very dry and fire weather extreme. Be aware of total fire ban days. It is important to have some readily available water accessible when using your smoker in the summer. A firefighting knapsack spray would be ideal but a pressurised garden sprayer is inexpensive and contains enough water to both extinguish your

*Begin lighting your smoker – you can start with some scrunched up paper*

*Gradually add fuel, keep puffing*

# CHAPTER SIX

*Do not over compress your fuel*

*Ensure it is alight, keep puffing*

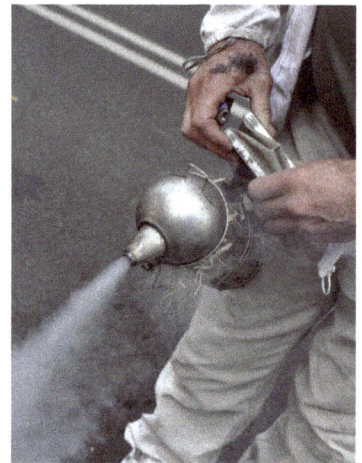

*Ensure it is alight, keep puffing*

smoker and provide an immediate response. 'Never extinguish your smoker by tipping it out on dry grass or the forest floor, empty it on mineral earth or gravel road and soak with water'.

Never leave a burning smoker in a vehicle.

Extracting

When using frames with foundation, the aim is to maintain the drawn comb for re-use by the bees. Wax costs a lot in terms of honey production, it takes the equivalent of 8kgs of honey to produce 1kg of wax, so conserving already drawn comb is important to commercial beekeeping.

Extracting Equipment:

If you have one or two hives the equipment can be borrowed, hired or shared. As the scale of operation increases, the volume of honey to extract and handle in a relatively short period of time will determine the equipment you need. Larger operations have motorised extractors, sumps, honey pumps and most will heat and cool their honey requiring specialised equipment. Some larger packers will have mechanised bottling and labeling systems.

• Uncapping knife or uncapping machine. A hand held uncapping knife, that is either heated by immersion in hot water or electrically. If using hotwater it is best to use two knives so one remains in the hot water keeping it hot. A scratcher comb is a small hand held device that is used to perforate or tear the wax cappings of comb that is in corners or sits below the frame edges and gets missed by the uncapping knife.

*Uncapping knives can be electric, steam or water heated.*

*A manual two frame tangential extractor for 1-2 hives*

*An electric 12 frame radial extractor*

- Extractor: from a simple hand cranked 2-4 frame model to a motorised radial extractor with no need for reversing frames. Except for the basic models based around a plastic bucket, they are made from stainless steel. It is important to ensure an even balance and when comb is full to begin extracting at a slow speed and build up the revs. Too fast and the centrifugal force will break the comb. Parallel extractors require the frames to be reversed whereas radial extractors do not.

*An industrial 80 frame radial extractor*

*New extractor technology*

- Cappings tank: Is a type of container with a coarse sieve able to suspend the wax allowing honey to drain. You can make your own and commercial models are available.

- Sieves and strainers with different mesh sizes to progressively strain honey removing wax, bees and bee parts, micro-mesh will remove some pollen.

- Containers, ranging from larger stainless steel drums, to food grade 20l plastic buckets.

- Wax melter, at its simplest is a stainless steel pot with a double boiler.

- A honey house or appropriate equivalent.

Bottling

- Honey gates, a shut off valve that stops honey flow, almost drip free!

- Settling containers, to allow trapped air and any remaining solid contaminants to rise to the surface.

- Retail containers ranging from self-sealing plastic buckets to glass jars

- Labels, marketing and a food label.

Honey House

Honey that is for sale needs to be extracted at a licensed food premises such as a commercial kitchen. It is possible to have your home kitchen licensed for limited uses within most municipalities in Tasmania. Of course scale is important, a bumper crop and you'll wish you never heard of bees and honey if you are extracting 100's of kilos in your kitchen. A model honey house like that shown supports an apiary of 25 – 30 hives and is more like a laboratory.

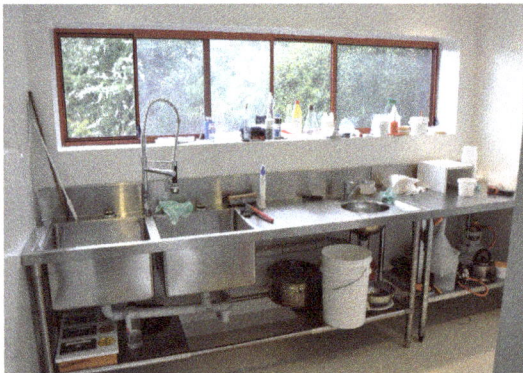

*A great example of a small honey house (EEee's Bees)*

*An industrial honey house (Tasmanian Honey Company)*

Stainless steel benches, deep industrial sinks, wash down areas, good storage and above all a clean, hygienic environment to extract and bottle honey, a model to aspire to and a credit to a great beekeeper.

## Refractometer

A refractometer is a device used by beekeepers and packers to measure moisture content. Designed to measure the refraction of light through a liquid and calibrated to provide a moisture content range. Honey should not be stored for sale unless the moisture content is less than 18%. Honey that is not pasteurised, raw honey, contains natural yeasts that are active in moisture contents greater than 20% and will eventually cause fermentation. Less than 20% the high sugar concentration causes

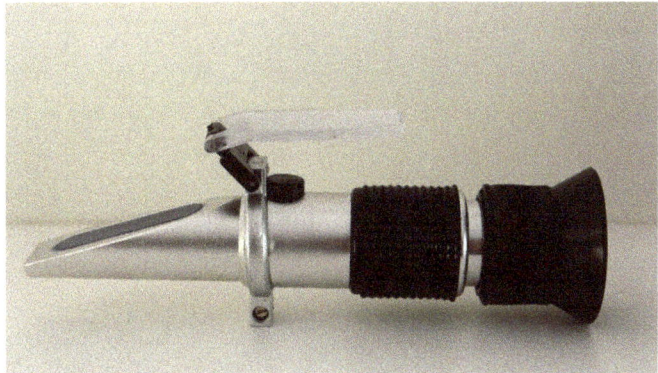

*Moisture refractometer*

the yeasts to become dormant, giving a much longer shelf life. Detailed instructions come with the devices, you first calibrate the device using distilled water then you smear a thin honey sample on the prism, look to the light source and read the scale. You must ensure that honey is mixed well before sampling and take an average of at least three readings. This will provide you with some confidence, but remember honey is a natural product and will vary across samples unless it has been pasteurized.

*Maintaining your equipment is essential, keeping it clean and disease free is a key to your colonies health.*

## Equipment hygiene

Maintaining your equipment is essential, keeping it clean and disease free is a key to your colonies health. Many disease and health issues in bees are caused by beekeepers and the negligence of hygienic practice contributes to this. Let us practice what we preach, if we want hygienic queens, producing bees with hygienic traits, then let it begin with the beekeeper.

Always sterilise your hive tool between visits and wash your protective clothing. Leather gloves require a gentle hand wash with a mild detergent and once dry, the application of a non-perfumed leather treatment

Hive Registration

Hive registration will be compulsory and is recommended that all beekeepers no matter how small register their hives. This provides a data base of registered hive locations, an essential biosecurity measure. No one wants a disease incursion that can decimate the industry and knowledge of known hive sites, especially those of amateur beekeepers is very useful for education and prevention of disease spread. The TBA Code of Practice for Urban Beekeeping specifies that urban hives are to be registered.

A registration form is available for download from Biosecurity Tasmania.

Equipment Supplies

There are quite a few Australian stockists of beekeeping equipment. A list of popular suppliers is appended that starts with your local Tasmanian suppliers. It is important to consider them first, if we do not buy some gear from them, when you're desperate it may not be available or they will have stopped the bee side of their business. Ebay is always an attraction and there are some good buys but beware of cheap imports.

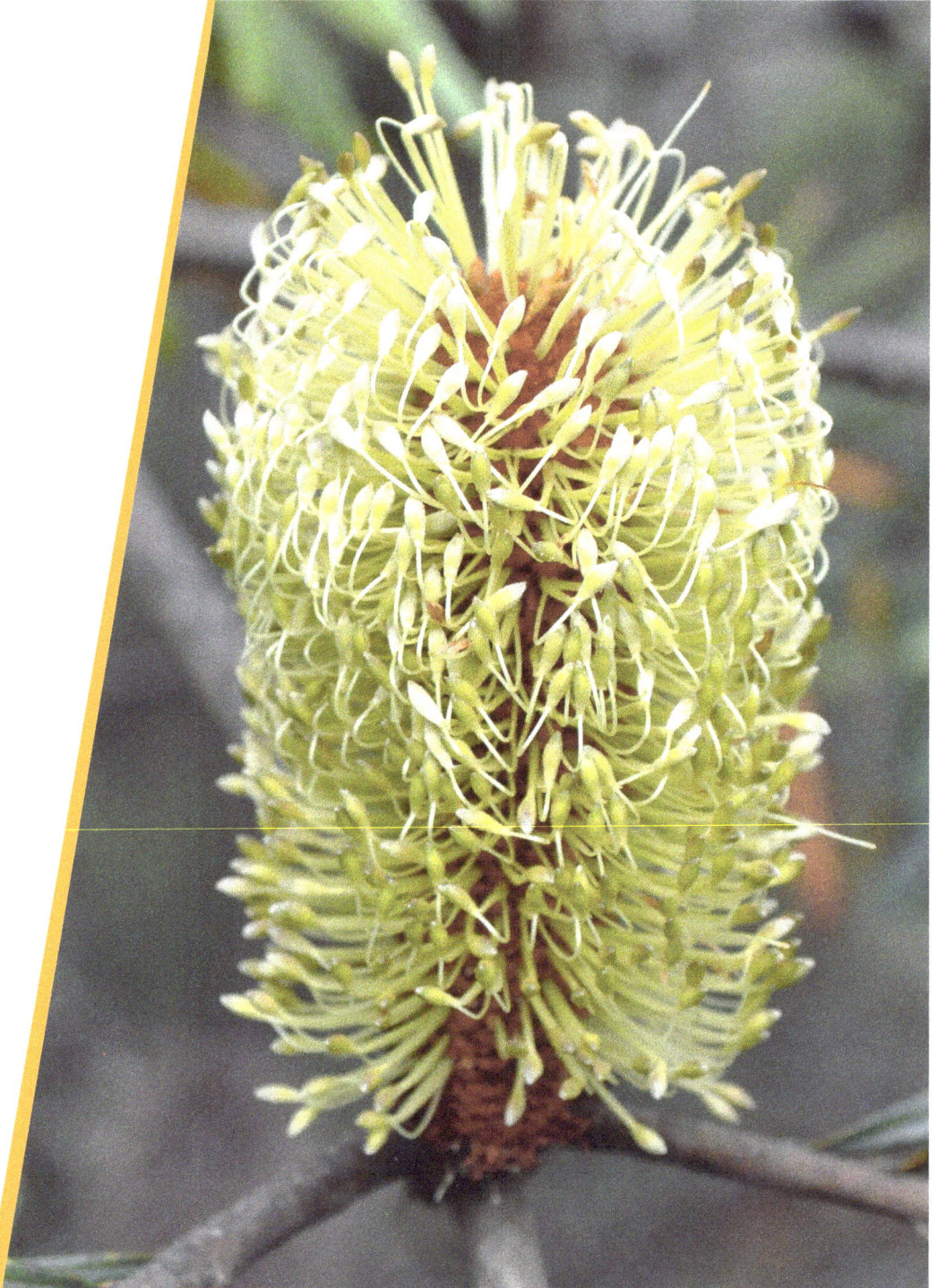

Keeping a bee journal is a very useful habit to cultivate and will help you become aware of different issues, requirements and events in the hive. You can record things you see in your hive as well as the date of flowering events.

The means of recording can be anything from a small notebook to a ring binder or a downloadable application for your smart phone or tablet. Use of a smartphone or tablet during hive inspections may prove difficult with honey soaked gloves. However the technology has an amazing capacity to store, compare, and track hive management.

Recording information about colony condition and activity will encourage you to take care of the basic requirements and is a key to successful beekeeping. It will help you from repeating mistakes and predict future events and hive condition.

Start your records from the day you establish your hive noting the date of each visit, the weather on the day, what is flowering at the time and what you see in the hive. The following may provide a useful checklist:

- Location of the hive and directions to get there.

- Date of visits.

- Weather conditions, current temperature, whether it is sunny, cloudy, still or windy.

- What is flowering, carefully observe as it may be a pasture species like clover or dandelion as well as more obvious shrubs and trees.

- Weather between visits: rainy, dry, hot or cold?

- Colony strength, the number of frames with sealed brood and the number of frames covered by adult bees.

- Colony characteristics, aggressive, docile, productive or hygienic.

- What you observe in comb: eggs, larvae, capped brood, brood pattern, honey, pollen (colour) anything unusual, any sickness or disease.

- Number of eggs, open and sealed brood in a 1:2:4 ratio (Flottum:2010).

- Signs of swarming: queen cups, queen cells.

- Manipulations made: reversing, supering, looked for queen and found her or didn't find the marked one you bought.

- Amount of honey robbed, number of frames and where from or the number of full supers.

- Diseases of the brood and pests. See chapter seventeen Diseases and Pests.

This may all seem unnecessary but as you build your skills base you will find that reflecting on your past actions has been very helpful in guiding your management.

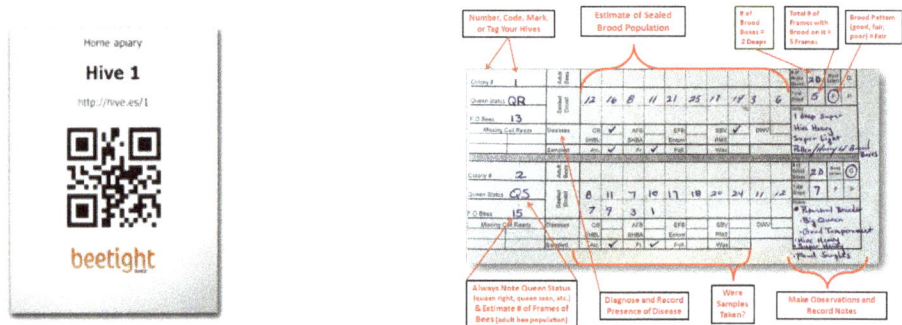

Beetight a smartphone app is able to produce QRT labels that provide instant access to hive records.

An example of a hand written field note from the Bee Informed Partnership US

After some years you will most likely reduce this to keeping basic information; swarming, queen information, breeder and production.

As a minimum record:

- Disease, pest, biosecurity information.

- Hive movements.

- Queens; the breeding of your queens and their age.

## BIOSECURITY

Keeping Australia free of exotic bee pests and early detection is every beekeepers responsibility regardless of your size, one hive or five thousand.

Keeping accurate and complete records of biosecurity actions should be done. If you detect or suspect an exotic disease or pest call the Exotic Pest Hotline.

The Biosecurity Manual for Beekeepers 2016 provides complete information on biosecurity observation and record keeping and is downloadable from their website. More detail on existing and exotic disease is covered on the Chapter Diseases & Pests.

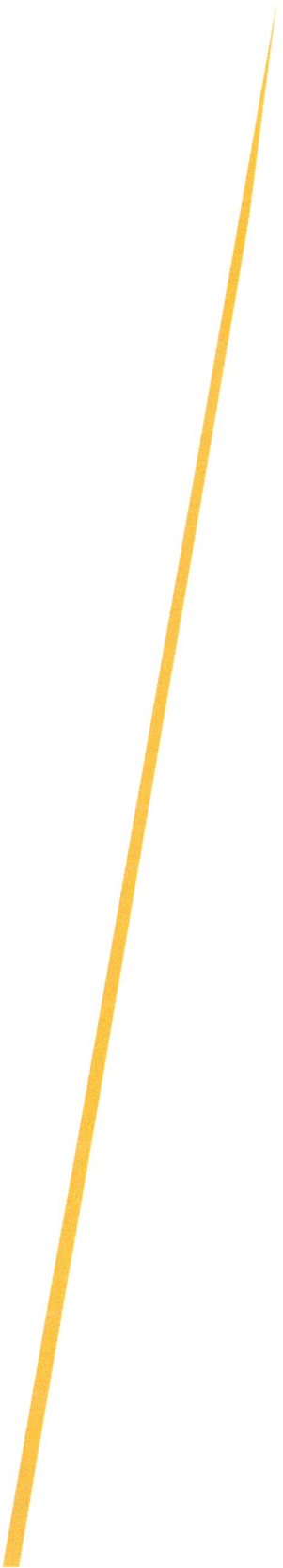

Bee stings are an occupational hazard. Bees have a barbed lancet or sting that can penetrate skin and cause a variable reaction in the individual from little or no pain to life threatening anaphylactic shock.

For most people bee stings are painful resulting in a local reaction of pain, swelling and itchiness. This location reaction is a mild reaction that can last a few minutes to a day. For some people the reactions can increase in severity to extreme levels requiring immediate medical attention. The stinging bee, a female worker has just lost her life defending herself and or the colony. As they struggle to fly away the poison sac is torn out of their abdomen meaning ultimate death and the inability to re-sting. Worker bees can repeatedly sting other bees, their barbs only get stuck in leathery hide or skin. Queen bees have a smooth lancelet (stinger) and can repeatedly sting, mostly used against other queens, rarely against the beekeeper.

*This digital composite depicts the barbed lancet stuck in a persons skin "stung" the end of the worker bees life! Digital Composite Marc LeRoux*

*Bee venom at the tip of the lancet of a worker bee. Zachary Huang*

The worker bees lancet is barbed, locks into the skin and acts as a self-injecting mechanism pumping more venom after detaching from the bee. It is important to remove the sting as fast as possible. When stung, scrape the sting off with a fingernail or hive tool as soon as possible, reducing the amount of venom pumped into the wound. Avoid squeezing between your thumb and fingers as this pumps more venom into the wound before removal. Speed is more important than technique, quick removal of the sting reduces the amount of venom injected and reduces the amount of swelling and itching.

An alarm pheromone is released with the sting and will attract other bees, with potential for more stings! It is good practice to apply smoke to the area to mask the pheromone.

*Anaphylaxis is a severe allergic reaction and potentially life threatening. It should always be treated as a medical emergency, requiring immediate treatment.*

Allergic reactions

Anaphylaxis is a severe allergic reaction and potentially life threatening. It should always be treated as a medical emergency, requiring immediate treatment. Anaphylaxis first aid has been provided by the Australian Society of Clinical Immunology and Allergy.

Symptoms can be mild to severe from facial swelling, tingling or stomach pain to difficulty breathing, swelling of the tongue, dizziness and collapse. This must be treated as a medical emergency. If the person is conscious they should self-administer the auto-injector. In severe reactions where the person is incapable of self-administering, lay the person flat or allow to sit, administer adrenaline if they have an EpiPen® (adrenaline auto-injector) and make sure you count to 10 before removal, and call emergency (000 in Australia).

Used with permission of the ASCIA 2014.

- Prevention is better than cure, avoiding stings is not only safer but makes your beekeeping a more pleasant experience.

- Avoid extensive examinations in cold or poor weather as the older more aggressive bees are at home.

- Always wear protective clothing: a veil, veiled jacket or bee suit, gloves (many commercial beekeepers do not wear gloves), and gaiters over your trousers.

- A smoker is essential to have at hand, that has been well loaded and is alight. Do not hesitate to use it and remember to give it an occasional puff to keep it alight.

- Stay calm and use gentle movements.

- Avoid using scented products in your hair or on your body while working bees. Do not use personal insect repellents.

- Approach your hives from the side or the back, do not stand in front of the hive entrance.

If you are repeatedly stung

- Walk calmly and quickly away from the apiary.

- If there are trees or shrubs nearby walk amongst them.

- Get into your vehicle or a building if possible. Adapted from (Bee AgSkills 2013).

Bees can be bred for quiet behaviour and this is a trait you should ask for in any queens or bees you buy.

# CHAPTER NINE

Swarming and Swarm Prevention

Swarming is the natural means for bee colonies to reproduce and the population to grow, it is normal. But if your bees swarm and you are unable to recapture them you have lost your queen and half your bees reducing your honey capturing capacity.

Swarming in managed urban hives can be a problem to neighbours and is very alarming to the general public, and the cause of complaints against urban beekeeping. For beekeepers it significantly reduces the size of your hive approaching a honey harvest and may result in no surplus honey. The first or top swarm is usually the old queen who leaves the colony with half the workers, leaving behind a ripe queen cell to replace her, creating two colonies. This can be followed by other swarms or casts, usually much smaller and with a virgin queen.

It is important that you become familiar with the reasons for swarming and signs of swarm preparation and the practices to reduce it.

Honey bee colonies swarm for many different reasons including one or more of the following:

*A typical first swarm within a few hours of leaving the hive. J. Elder*

- Not enough room: This can occur in spring with early warm weather, plenty of food and a rapid build, then a period of poor weather and overcrowding occurs.

- A failing queen: This can be age related as the production of queen pheromone is reduced by half each year. Queen pheromone indicates the queen is present and doing her job and inhibits swarm preparation. Larger numbers of bees cause the queen pheromone to be spread too thinly as it is spread from bee to bee reducing its effectiveness. Instead of superseding, the colony may swarm.

- A light honey supply and abundant pollen: These conditions can occur in early spring and cause rapid expansion of the hive. If these conditions persist it adds to congestion with inactive worker bees and abundant brood (Somerville:1999).

- The genetics of the colony or the race of bees: Some are prone to swarming.

- Starvation.

- The colony is over stressed by too much beekeeper interference,

- Poor ventilation and or fumes from freshly painted boxes (oil based paint is more of an issue than acrylic paint).

Signs of Swarm Preparation.

- Queen cell or cup construction along the bottom edges of the frames.

- The queen ceases laying eggs after the production of queen cells, is restless and her abdomen reduces in size allowing her to fly.

- No room for expansion of the brood nest.

- A large worker bee population from favourable early spring weather and a minor honey flow.

- Field bees are less active and congregating at the hive entrance.

*Queen cell constructed due to lack of room*

*Regular inspection during in swarming season is essential. Removal of queen cells is a temporary answer, more room must be made*

When, not if it happens.

- If some or all of the above signs are present, then the hive will likely swarm on the first warm, sunny, calm day after prolonged unsettled inclement weather as the hive has become too congested.

- This usually happens between 9am and 1pm.

Prevention: It is best to keep your bees.

*Swarming may not always be prevented, but the occurrence can be significantly reduced by being diligent and following some simple procedures*

Swarming may not always be prevented, but the occurrence can be significantly reduced by being diligent and following some simple procedures, including:

- Re-queening regularly with a reduced swarming strain.

- Making space in a brood nest full of brood by moving frames of disease free sealed brood to weaker hives, fill the gaps with empty worker comb.

- Remove combs of honey and replacing them with empty combs.

- Spread sealed brood into a higher box with no queen excluder and place empty worker comb in the brood chamber, this gives the queen more laying room. Do not isolate brood as it will promote queen cell production.

- Inspect brood comb for any disease prior to removing frames of sealed brood and transferring them to weaker colonies.

- Commercial beekeepers constantly move brood frames, seasonally grading and removing old discoloured brood comb.

- Take care with moving brood early in the season as it can lead to chill and the development of chalk brood.

## Splits and Artificial Swarming:

The swarm tendency and colony overcrowding can be reduced by either splitting the hive in half, creating two new hives of equal strength or creating one or more nucleus hives from a strong colony. A nucleus hive or nuc is made of three to five frames by removing some brood frames containing equal eggs, uncapped larvae, capped brood, enough younger nurse bees to

*Beekeepers increase the size of their apiary by inducing swarming, creating nucleus hives. Removing brood and adding queens release pressure in the main hive and creates a new 'mini-hive' or nuc*

cover the brood and two frames of honey and pollen stores to provide initial food. It is important to ensure the queen from the main colony is not on the brood frames that you transfer. Common practice is to introduce a newly mated queen to the nuc. The nuc can either be recombined with the hive later in the season when you would kill the old queen, creating a strong honey producing hive, or you have just increased the size of your apiary.

Another very simple method of reducing colony size and creating a split is to remove the brood chamber to another position, place a new chamber on the original site with the queen on a frame of open brood with some honey in the middle of the super, this will help them stay. In a strong honey flow the box can be filled with frames with foundation that will be drawn very well. The field bees will come back to the new box and it will act like a swarm rapidly building new comb as the queen needs cells for laying eggs.

A new queen can be introduced to the parent hive or you can look for queen cells, choosing the two best formed ones and destroying the rest.

It is important that you take particular care in the spring to reduce this problem and remove swarms promptly from your neighbourhood. All beekeepers have a responsibility to know how to remove a swarm.

Swarmed bees are most ready to draw comb to provide storage and cells for the queen to lay eggs and are often used for this purpose.

Collecting a Swarm

The top, prime or first swarm headed by the old queen will initially settle close to the hive, often in a branch at about head height. It is held together in a football sized formation by worker bees producing a pheromone from their Nasonov gland.

*Preparing for swarm capture, a super sealed at the bottom, a lid and a smoker*

*A very small secondary swarm that caused concern for the residents*

If you are quick in responding to this, within 1-2 hours you may be able to easily collect this swarm by one of the methods explained below. If left for any longer they will most likely go to a pre-determined new home located by scout bees as a safe, sheltered site where they can build comb and the queen can start laying. The latter is a more difficult recovery operation.

The most suitable recovery method will be determined by where the swarm has clustered.

*A successful swarm capture*

If you spot the queen in a swarm and are able to catch her, bonus! Place her in a hive and the bees will follow. Otherwise here are two simple methods that have been successful, and there are others.

You can use any box if you are caught out, but plan for swarm capture with a super with a base, top and one or two drawn frames of dry comb. Make sure you have your regular equipment

- Bee suit and gloves.

- Smoker, fuel and lighter.

- A bee brush to brush bees into the box.

Shake or place the swarm into the box (lid off and bottom board in place), if it's a branch and can be shaken, if too high ask for permission to cut it then shake either over or in front of the box. Leave the box on location until evening, then reposition in your apiary.

Once the bees are in their permanent home or super, add additional frames of foundation or drawn dry comb. Adding a frame of disease free brood from another hive will increase the chances of the swarm staying. To replace the queen, allow two to three weeks to ensure there is a queen and that she is laying. A prime swarm will usually produce a supersedure in about eight weeks.

An alternative is to place the hive above the swarm with the top on and bottom board off and gently smoke the bees into the hive, replacing the bottom board and leaving in place until evening, then repeating the previous process.

*To ensure you do not introduce disease, hive the swarm on foundation. They will draw comb, consuming the honey they have bought with them instead of storing it*

To ensure you do not introduce disease, hive the swarm on foundation. They will draw comb, consuming the honey they have bought with them instead of storing it. Keep the newly housed swarm separate from your other hives for two to three months.

If you are a beginner, it is advisable to take an experienced beekeeper with you or to accompany an experienced beekeeper when they undertake swarm collection.

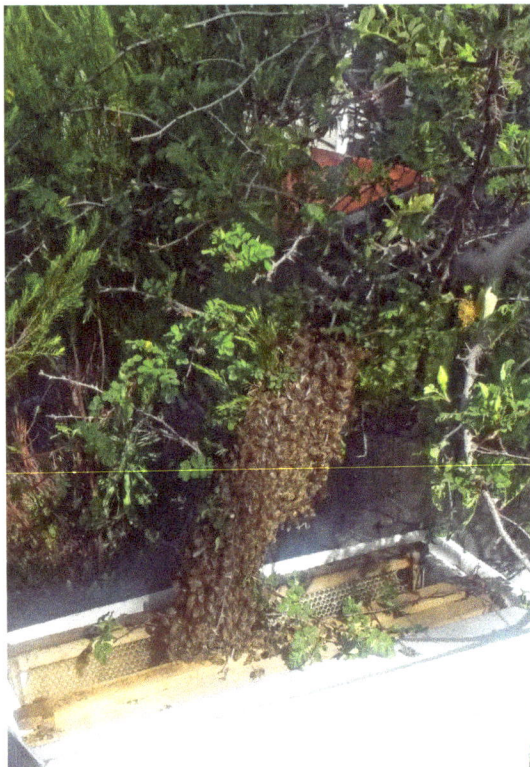

*Shaking a first swarm into a nucleus hive. Melbourne City Rooftop Honey*

# CHAPTER TEN

Hive Inspection and Bee Handling

Inspecting your hive is something that will become second nature, but first you need to overcome your anxiety, even excitement and plan your first visit. This is a simple step by step overview of what you need to consider, what to take, what to look for and to avoid.

*First and always, be prepared, make a list of what you need, keep it in your box or container of equipment and check it off.*

First and always, be prepared, make a list of what you need, keep it in your box or container of equipment and check it off. While it may become second nature many experienced beekeepers have found themselves on site without their lighter, fuel, smoker, hive tool or even their protective clothing. Also it is essential to let someone know where you are going, preferably with a map or GPS coordinates of your apiary location/s and when you expect to return. If you are a beginner it is advisable to practice lighting your smoker and ensure you can keep it going and to understand the use of different fuels. The Equipment chapter six has a step by step pictorial of lighting your smoker.

On the day of your visit do not use strong smelling deodorants, shampoo, other cosmetics or insect repellent, they definitely upset bees. It also pays not to wear exposed woollen or dark clothing and not to have had contact working the horses or patting your dog, even eating a banana on the day (bananas contain the same chemical as the alarm pheromone).

Once at your hive site don your protective clothing .

- Make sure your gloves cover your sleeves.

- Your gaiters cover your ankles.

- Your bee suit is fully zipped.

Pick your weather, a fine sunny calm day is best, avoid wind and cold temperatures as the forager bees will be in the hive and are the more aggressive bees. When the weather is fine and above 13°C or more, the foragers are out working, the hive population is reduced and you'll find it a more peaceful experience.

*Fine weather and preferably 17° C or more for the first spring visit.*

*Unsettled bees after inspection.*

If it is in your backyard, warm and sunny during weekends or holidays it will usually mean your neighbours will be outside. Attend to your bees in the hottest part of the day, you'll have less trouble as the more aggressive workers are out on the job. If you inspect early or late in the day when there are more workers in the hive, unavoidable for commercial beekeepers, the colonies will take several hours to settle down once the hive has been opened and inspected.

With urban hives it is always good to keep on side with your neighbours and inspect when you will cause the least disturbance.

*With urban hives it is always good to keep on side with your neighbours and inspect when you will cause the least disturbance.*

Planning Inspections

Remember to plan your hive/apiary inspection before you leave home.

What is the reason for my visit and what will I be doing?

Do I have all the tools I need for the tasks I plan to do, including fuel and matches/lighter?

Is my bee suit clean, gloves and suitable footwear (not sandals or thongs)?

Is my smoker clean, with plenty of fuel, matches or lighter.

Do I have a spare container for burr comb and old wax and a smaller one for propolis?

Do I have cleaning materials to keep the hive tools clean and hygienic?

Make a list and check it off before leaving.

When on site have your smoker lit and ensure your bee suit is fully zipped up, the matches/lighter are in your pocket.

*BBKA 2012*

Spring Management

Before inspecting your hive, be prepared. This applies whether it is your first time or you are more experienced. Choose a warm, settled day preferably when temperatures are warming up, some flowering is occurring and bees are moving freely in and out of the hive. While it may be tempting to look at your hive earlier in the year, it is much better practice to minimise opening the hive in the winter, waiting until early spring and then only for a quick first inspection. With the onset of some warmer days and the beginning of flowering and food availability, the bees are probably raising brood and there is an increasing stress on the colony and its available food stores. Your first inspection is to determine quickly the condition of the hive and the availability of stored food.

Opening your hive is a stressor for the colony and added stress can be a precursor to disease.

If this is your second season, remember to look at your bee log or record book to see what condition the hive was in last time you looked. This is a good habit to form and act on what you see. As the season progresses this will become more important with the construction of queen cells and pre-swarming activity.

At your first post winter inspection on a warmer sunny day you are looking for the following:

Where are the bees?

- Before smoking and opening the hive, take some time to observe what the bees are doing. If they are returning with pollen it's a good indication that the queen is laying and brood is developing, the demand for pollen is increasing.

*Observing your hive for evidence of bee activity, dysentery (faeces on side of hive) and dead bees at the entrance.*

*A fine September day and the bees are active, little pollen in this image but pollen was coming in indicating expanding brood and an active queen*

- How much food is present, there should be at least 10kg of stores, equivalent to one full 'ideal' super, to provide for the growing colony while the weather improves and available forage increases.

- Look along the top of frames and top corners for sealed and partly filled honey cells, these are the last to be emptied. If there is no evidence of honey stores then the colony has run out of food.

- If you do not want to stimulate rapid increase in the hive, rather maintain it the colony will need to be fed.  Either feed a thick sugar syrup in an appropriate feeder or 2kg of dry white sugar placed on the inside cover this will provide enough food for the hive as spring flowering commences. While they may discard some of the dry sugar it will not overstimulate the queen.

*Good stores remaining in early spring*

*New brood with stores in corners*

*Absence of stores and chewed corners indicates food stress*

- Feeding thin syrup will tend to stimulate the queen to start laying, increasing your risk of overcrowding and early swarming. This method of early hive stimulation is used by commercial beekeepers engaged in pollination of early flowering crops.

- Clean off bottom boards, wax cappings and dead bees and note which hives had few dead bees and overwintered well. Remove the waste from the site.

- This should be a quick look trying not to disturb the bees more than necessary.

As this will be your first inspection of the season it is a good chance to remove propolis (bee glue) and all burr comb from between the supers, frames and lid, making the next inspection much easier as hive elements will not be stuck together.   It is also timely, although still early to create some more space for the queen. If they are over spaced remove boxes to bunch the bees up a bit.  You may find that the bottom super is nearly empty, move the brood box onto the newly cleaned bottom board and place the empty super on the top of the brood as bees and the queen move upward and it increases egg laying space, providing room for expansion. This will also help prevent swarming. This is also a good time to begin the process of removing old discoloured comb and replace with good drawn comb or foundation.

The second inspection post winter is your first major inspection and is more thorough 1-2 weeks after the first visit depending on the weather. This is done when the weather is generally hotter than 17°C. Look at the forecast selecting for the best weather and plan a day so that you are prepared. Check carefully through frames to ensure the hive is developing well.

*Significant propolis present at first spring visit.*

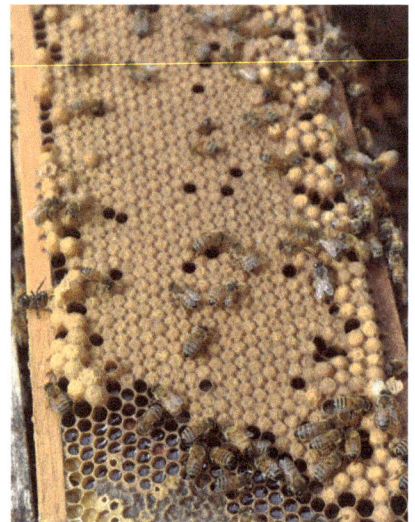

*Good spring brood development*

Look in the brood for the type of cells, worker, drone or queen cells as well as worker, drone queen cells mark their progress, open, sealed and just hatched, the quantity and pattern or distribution of the brood. Brood production normally commences in the centre of the comb and advances outwards eventually occupying at least six inside frames.

*An absence of brood may indicate an acute food shortage of pollen and honey stores, low temperatures or a dysfunctional queen.*

An absence of brood may indicate an acute food shortage of pollen and honey stores, low temperatures or a dysfunctional queen. Spotty or irregular brood with depressed cappings indicates dead brood and is a characteristic sign of brood diseases and you must investigate this further. Scattered brood or isolated groups containing workers and drones may indicate an aging queen. Groups or scattered cells containing only drones indicates a drone laying queen or the queen is dead. A drone laying queen will lay a single egg in the bottom of the cells whereas drone laying workers will lay multiple eggs all over the cell.

This is a good time to look for the queen in the brood comb as bee numbers will be reduced coming out of winter. If not already done, the queen can be marked and her wing clipped, but this can lead to supersedure. Note that wing clipping is not allowed in organic beekeeping.

The condition of the hive and the quantity of surplus stores can be checked by

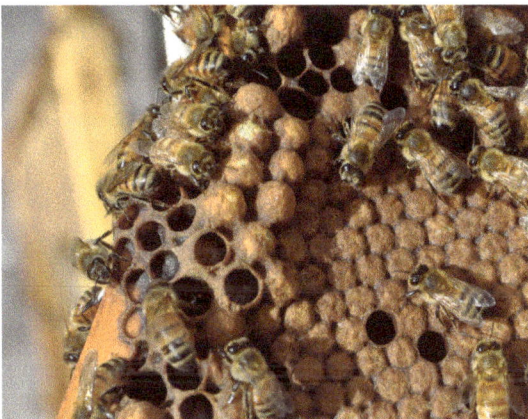

*Brood: worker (smaller), drone (larger domed), queen l(arger acorn shaped, unfinished)*

*A marked queen, easier to find but practice not accepted by organic beekeepers*

feeling the weight of the hive by tilting it. If you want to split the hive you can stimulate it with sugar syrup. If stores are reduced and you want to maintain the hive for an early honey crop feed thick sugar syrup, thin syrup stimulates the queen to lay.

If they come into spring with plenty of stores you can create room, even take excess honey, but this is not recommended as the weather and flowering can be very variable in Tasmania at this time of the year and the expanding population will need the extra honey and save you the work of supplementary feeding. Honey also acts as a heat bank as it warms up and helps maintain hive temperature.

If there is a lot of propolis on rebates and frame shoulders, remove it with your hive tool. You could bring a container along and collect it for processing later. Burr comb should also be removed as it makes future hive management difficult.

*A lid full of burr comb often resulting from lack of 'bee-space'.*

Look for bad comb; heavily darkened, sometimes warped in the bottom box and with corners chewed out, especially near the hive entrance. Move any empty dark comb to the outside of the box. In one month, pull out and replace with fresh foundation or drawn comb. Having a brood comb replacement program removing old dark, small cell comb helps maintain cell size that equates to bee size. Small cells produce small bees. Cells reduce in size as wax cappings and other waste material from pupation builds up in the cells reducing cell size.

Queen cells if observed need to be taken out. If queen cells are present at this early stage hive observations need to be made every eight days to remove queen cells and help prevent swarming. This is also time to take a nucleus out creating a split hive, this is called induced swarming.

*Queen cells in swarming season*

*Queen cells removed with hive tool*

> *As the season advances quick regular inspections should be made to ensure the bees have plenty of room for storage of pollen, nectar and expansion.*

As the season advances quick regular inspections should be made to ensure the bees have plenty of room for storage of pollen, nectar and expansion. Swarming behaviour can be induced by lack of room, but generally does not happen in Tasmania until late October.

As the season moves into summer, regularly look for queen cells and remove them. If you find a 'ripe' but empty queen cell, the hive has probably already swarmed. You should check again in 10-24 days to see if the new queen is laying.

Splits

Splitting over populated hives is a form of induced swarming and is used to reduce the population of one hive by creating two hives with smaller populations. These new hives will still need to be observed for swarming tendencies. If the weather remains favourable and there is a strong honey flow you may need to put at least one empty super on.

There are a number of ways to create splits, the following is a simple method most often used and is suited to urban situations as you do not increase the number of "boxes". The key that requires some experience is to maintain the right balance of adult bees and brood. In the morning pull brood about to hatch and stores and place in another super along with some nurse bees. Place some comb or foundation in the empty spaces in the bottom box then place a queen excluder between the bottom super and newly populated one, of course with a lid on top. Nurse bees will move up through the queen excluder to attend to the brood.

The evening of the same day remove the queen excluder and replace with a division board with the entrance facing in a different direction to the bottom one. Any foragers will leave and go back to the bottom box leaving young bees in the top box. Insert a caged laying queen or a capped queen cell into the top queenless super. If a capped queen cell check after 12 days and 7 to 10 days after a laying queen has been introduced. The top hive will be fairly quiet for about 3 weeks but then foragers will begin to emerge and the cycle continues (Hewitt 2016).

Strengthening Weaker Colonies

Your aim is to have strong colonies able to optimise the honey harvest. If you have weaker colonies they can be united with a stronger colony. You first need to determine the reason that the colony is weak. Is it queenless, is the queen old or does the hive have a disease? If diseased you do not want to unite it with your stronger hive. Before uniting you will need to find the queen in the weaker colony and kill her.

The easiest method to unite two hives is one using newspaper to separate the hive bodies as the bees from the separate hives will fight if combined immediately without time to adjust. Open the strong hive and place a sheet of newspaper over

*Division board with a small entrance*

*Newspaper combining*

the top bars covering all frames. Some small slits in the newspaper may help but many commercial beekeepers don't bother and it works well. Lift the weaker hive from its floor and place on top of the newspaper. Leave the hive for several days before checking that it has successfully united. As it takes a day or two for the bees to chew through the paper, the two colonies adjust.

There are other methods, but this is the most successful and recommended for beginning beekeepers

Transport

*Eaten out newspaper. Holes chewed out overnight, 1-2 days to eat out*

Transporting your hives is a serious undertaking that requires some specific approaches and appropriate care. Prepare for an emergency, know who to contact and what to do in case of an accident. You are carting 1,000's of potentially stinging creatures through populated areas. The bees cannot (should not) escape and will be under some stress no matter how well you undertake the operation.

The best time to close hives is around dusk when bees are no longer flying, all the foragers have returned. Many large operators will then load their trucks ready for a 4-5am start allowing them to unload in daylight before temperatures rise. This may not be the case in hotter parts of Australia or the world where moving your hives early in the morning especially early summer is not advisable, as temperatures soon increase placing additional stress on the colonies. The Australian Honey Bee Industry Council AHBIC, has developed a Best Management Practice for the Transportation of Open Entrance Hives.

*An early morning trip to the south west leatherwood forests. A well secured load*

*Large loads of bees must be well secured*

This emphasises the need to transport at night, to be prepared, secure your load to standard, not to stop near lights and preferably not to stop on your journey.

Summer movement:

In Tasmania this relates to moving the majority of commercial hives and some amateur hives to leatherwood sites in the rainforests of the northwest, west, southwest and southern forests. It is advisable to remove most honey supers before moving to reduce weight and extract them as soon as possible. One to two frames of honey are often left on in case you arrive in wet weather. Move hives with eight ideals or four full depths providing plenty of room during transport and for a potential heavy flow. If you have just been on a prickly box (Bursaria spinose) flow this does need to be extracted as it will set in the comb. If the weather is forecast for higher temperatures it is very important to spray water into the hives every 1 to 1.5 hrs. A pressurised garden sprayer is adequate, stop your vehicle and attend to this as overheating during transport can kill bees, even under nets this can still be a problem. Moving to a remote location where the honey flow can be potentially rapid and significant requires plenty of room for honey storage, initially four ideal honey supers or two full depths.

Hives will be transported for a number of reasons; moving in late autumn from the last honey flow to your winter/spring sites, moving splits and for pollination and honey flows when the populations are at close to peak numbers. Moving splits with small bee populations in spring must still be done with care to ventilate the hive if travelling more than 1hr. In high population moves in late spring and

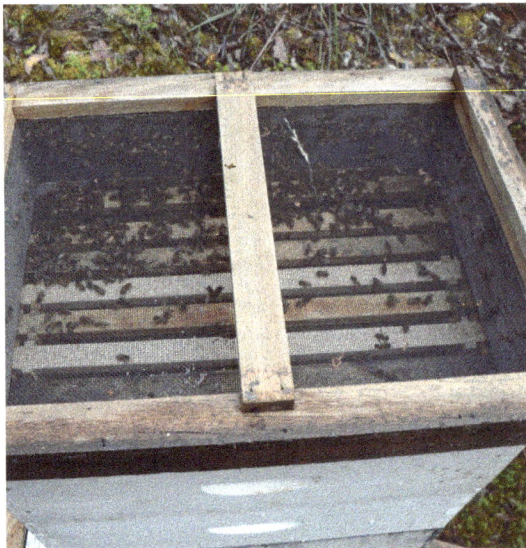

*Bees after a 2.5hr drive filling the transport screen*

*Bees in the lid above a perforated metal spacer*

summer heat generation within the hive is high.

When about to close your hive entrance as the bees have stopped flying, a few light puffs of smoke at the hive entrance drives any stragglers into the hive. As hive entrances are closed or even under nets the bees need more room. At least an empty super or half super on top and a transport screen or some other form of ventilation is essential. Hold your hand at the top of a closed hive or on the metal spacer-vent and you will notice how much heat is generated by cold blooded insects!

The top screens that provide 50 to 75mm clustering space are ideal and readily made from flywire and timber frames with cleats to lift the lid above the frame as in the image. Netting loads with open entrances is the preferred, efficient means for commercial beekeepers moving whole truck or semi-trailer loads, usually done at night.

It is important to secure your load whether 1 or 100 hives. Place hives side by side, limiting hive movement, ensuring ventilation, with frames running parallel to the line of travel. Commercial operations usually have the hives on pallets, regardless of configuration all loads must be secured. Modern webbing ratchet straps enable loads to be well secured.

Inspect the hives after travelling 10km, make sure the load is secure!

*Removing shade cloth from open-entrance transport*

*Larger commercial beekeeper unloading pallets of hives.*

Travel at night or early in the morning and if the hives are on a truck, keep the engine running when you are stopped, this keeps the bees clustered and less frenzied also reducing overheating and remember to use a cooling water spray if temperatures are high.

Set your hives out on their new dry site so they get optimum morning sun with light afternoon shade, tilt them slightly forward to shed any water, have enough space between hives for ease of working, slightly different angles of orientation or colour can help prevent drift. Vehicle access close to the hives is most important in all types of weather.

Once hives have been re-established on their new site remove the entrance blocker as soon as possible, if its daylight, a few puffs into the entrance as you remove the entrance blocker will slow the bees down while you attend to the rest of your hives.

*Unloading bees at a leatherwood site*

*Apiary identification is a legal requirement*

Check that your hives have been stencilled with your surname and usual place of residence. The Animal Health Act 1995 requires 1 in 10 hives to be marked with the owners contact details.

Supering

Remember one hundred different beekeepers, one hundred different ways to keep bees. Supering is one of those subjects where people differ in terms of super size and supering. In Tasmania the majority of commercial and amateur beekeepers use ideal supers for honey, some use full depths of brood and some only use ideals for all their operation providing ready matching of all hive elements. Ideals are more convenient weighing about 10kgs when full of honey.

*Ideal supers being used for brood and honey*

*Full depth supers used for brood and ideals for honey*

However, it is contested by some commercial producers that full depth hives are more efficient in terms of handling and better managed by the bees.

Now it is time for honey collection, the bees need room to store honey. Placing supers with drawn comb or stickies on top of the brood is a sure way of getting rapid storage of honey in a strong nectar flow. Adding extra supers is often done on top of the existing honey supers. If the hive is some distance from your base extra supers can be added at establishment. Placing a lot of foundation on top is not advisable as the bees will only draw it out in a very strong flow.

Indications of a full hive needing more supers are bees hanging at the entrance or the lid is full of bees when you remove it. It is good practice to remove the supers when they are full and replace them with supers of drawn comb. If you do not remove the honey you can stop the bees from working. If you can extract infield, putting stickies back on will mean you need less spare drawn comb.

You need to be prepared, even Prickly Box has been known to yield in excess of 100kg/hive in a very short time and leatherwood has exceeded 150kg/hive on the odd occasion but often yields more than 100kg/hive. With ideal supers holding about 10kg, this is either a lot of woodware or repeated regular robbing. The frequency of robbing will depend on the strength of the flow, with extreme flows requiring weekly robbing and an average of a monthly inspection.

*An over-supered hive in anticipation of a heavy leatherwood flow.*

*A regular inspection watched by a 'future' beekeeper*

If weather conditions in spring commence as cold and variable it is advisable not to oversuper until the weather improves. Early summer weather can also vary from winter light conditions with strong winds, rain and colder temperatures. This can change overnight to temperatures in the high 20's. Leatherwood can produce significant volumes in a short time, it is best to be prepared.

Honey harvesting or robbing should occur on a sunny warm morning when the bees are working vigorously and there is little fresh higher moisture content honey in the hive from the days foraging. Bees can be shaken from supers or clearer boards can be used if you are able to revisit in one or two days. If it is a very strong flow standing honey supers off to the side of the hive is effective as the bees return to the hive. When all the honey supers and frames have been removed fresh supers and frames can be added.

## Autumn and Winter Preparation

As the honey flows have completed it is time to check the stores for winter and into the early spring flowering. As a rule of thumb in Tasmania two ideal supers or a full depth of stores should remain on the brood chamber consisting of a full depth or two ideals. This will provide 15-20kg of stores enough to support the colony from May through to September. Do not put stickies on for cleaning before storing. Stickies help prevent wax moth infestation if you do not have access to a cool room. Some smaller beekeepers place a sheet of newspaper between each super being stored.

Only strong colonies should be overwintered as they will respond fastest in spring. As with spring management weak disease free colonies can be united with stronger ones going into winter. As the temperature begins to drop below 18°C bees will begin to cluster between the combs near the bottom of the stored honey. Bees need room to cluster, if the top super is full of honey, make some room in the centre by bringing up empty comb from the bottom. They will gradually eat the stores moving up the comb. Bees naturally make room by dying.

As you move to your winter sites avoid jarring the hives, protecting the cluster. Ensure hive components that need repair over the winter have been replaced and hive tops are water proof. The bottom board should also be cleaned before going into winter shutdown. The hive entrance should have been reduced in size going into winter. This also helps with irregular cold spring weather and colonies defend attack from wasps and a weaker hive defend the entrance against robber bees. A simple wooden mouse guard can also serve the same purpose.

The winter site should be sheltered from cold prevailing winds, a sunny aspect, not flood prone, preferably accessible in all-weather. A quick check of the hive to see if there is stored honey in the two outside frames and along the tops of the other six. This ensures there is sufficient food for about four weeks.  If this is absent feeding should commence as soon as possible. Hive entrances should be reduced in size by using a strip of wood or a mouse guard.

Bees in Tasmania do not usually die of cold, they starve.  Signs of starvation are

*An excellent winter site, close to the beekeepers home, all weather access, sunny and sheltered.*

*A simple syrup feeder from an inverted tin.*

*Bees in Tasmania do not usually die of cold, they starve.*

dead bees at the entrance and just inside the entrance. This will progressively increase, but as long as there is life there is hope, it may not be too late. In this emergency situation, the bees can be sprayed with a warm sugar or honey solution. Do not burn the sugar as it will cause dysentery in the bees.

Thicker sugar solutions or dry sugar, fondant or icing should be fed to bees in winter as you do not want to risk inducing brood production. Honey should be disease free and only use white sugar as brown sugar and molasses causes dysentery.

Feeders are available from apiarist suppliers or can be simply made from a covered can with a series of small holes punched in it. This is inverted over the top of frames inside an empty super. There are many variations, another homegrown example is to use an empty super containing a kitty litter tray with up to 6l of thick syrup and with a float for the bees.

Winter is also the time to undertake routine maintenance and to build new hiveware.

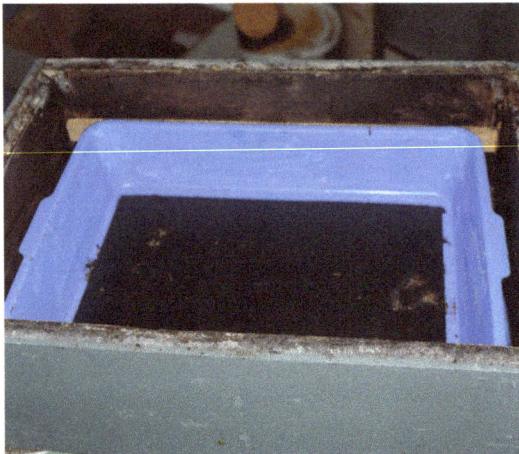

*An improvised feeder using a kitty litter tray*

*Commercial beekeepers need 4wd syrup tankers to ensure they can access hives in late wonter early spring when conditions may still be wet*

## Annual Activity Summary (not pollination)

### August

- Continue equipment repair and maintenance, order any remaining requirements
- Check stores by tilting
- First quick hive inspection only if weather is ok
- Order new queens

### September

- Continue emergency feeding if necessary
- 2nd. Inspection, thorough
- Clean out entrances and bottom boards
- Commence queen rearing

### October

Monitor stores especially if a cold, wet, windy spring.
- Unite weak colonies
- Rear queens
- Requeen split colonies
- Add supers to strong colonies where you may have Callistemon, Dandelion, Bluegum, Ironbark, Snowgum/Cabbage Gum or White Gum or some ornamental/ exotic plating or crop

### November

- Add more supers if needed
- Watch diligently for queen cells, remove
- Make space
- Capture swarms
- Rear queens
- Split strong hives
- Begin old brood comb replacement

### December

- Control swarming
- Capture swarms
- Draw foundation with newly housed swarms
- Remove spring honey
- Move to prickly box sites early to mid Dec

### January

- Remove honey
- Move to leatherwood
- Add enough supers

### February/March

- Monitor colonies
- Remove first take of honey and re-super by mid February to get a second take in March
- Final honey removal

### April/May

- Final honey extraction
- Prepare hive for winter
- Ensure enough stores
- Unite weak hives
- Move hives to winter sites

### June/July

- Check stores
- Do not open hive unless essential
- Repair and order equipment
- Order queens and nucs
- Attend conferences
- Go on holidays!!

# CHAPTER ELEVEN

Tasmania a large island 240km south of mainland Australia. It is situated in the middle of the Roaring Forties at latitude 42°South between the Pacific and Great Southern Oceans. It has a cool temperate maritime climate with four distinct seasons, warm to hot dry summers, and cold wet winters, with a windy and cold early spring, and often a mild very settled autumn.

The climate has a significant influence on the flora, its flowering time, productivity and availability to honey bees. Often when peak honey flows are expected in summer there are windy, wet conditions limiting floral productivity and flying conditions for bees. Horticultural crops that flower early such as plums, apples and cherry varieties face significant issues of cold, wet, windy weather requiring specific management of hives for effective pollination.

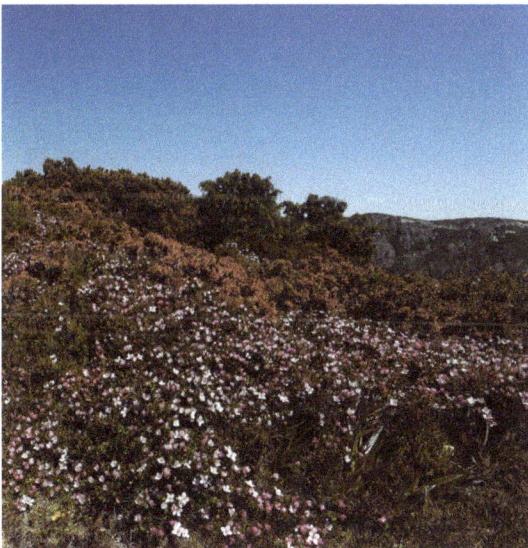

*Wildflower plains B Gibson (There are extensive areas of wildflower often underused)*

*Apple blossom provides valuable spring bee forage, especially pollen.*

Bees need flowers and flowers need honeybees and other pollinators for the success of their seed set, fruit quality and ultimately their survival. Managed honeybee colonies have become the preferred pollinators for many horticultural and agricultural crops and without them crop quality and quantity can be significantly affected. An often-quoted statistic is that one in every three mouthfuls of food we eat can be directly attributed to pollination by honeybees. Flowers provide bees with much needed food in pollen and nectar, essentially proteins and carbohydrates.

# CHAPTER ELEVEN

*Native and introduced flora contribute to bee nutrition and honey yield in an interlinked mosaic across the landscape.*

Native and introduced flora contribute to bee nutrition and honey yield in an interlinked mosaic across the landscape. (Leech 2012)

Honeybees and their colonies require diversity in their diet to build numbers and remain strong and healthy. A significant honey flow from one species may cause problems if the volume or quality of pollen is deficient and there is a lack of supporting flora. Some of our Eucalypt species such as Iron Bark, Eucalyptus sieberi, on the central east coast can be problematic, with occasional high nectar yields but a poorer quality pollen. Alternatively, some species can provide excellent pollen and an abundant flow enabling hives to build and provide a nectar surplus.

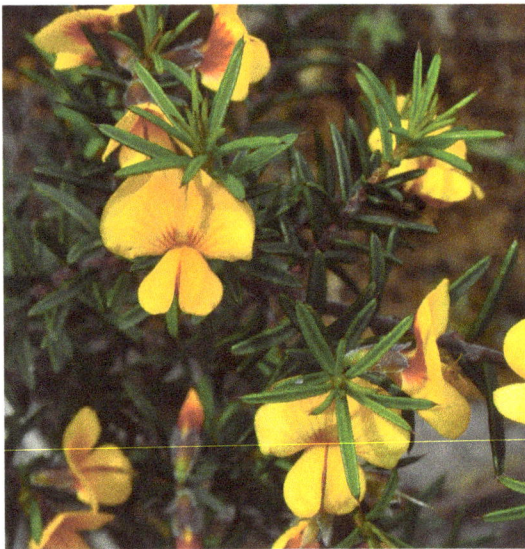

*Understory pea flowers are a source of highly nutritious pollen*

In the right conditions following a wet spring and moist soil to the time of flowering in December, Prickly Box, Bursaria spinosa can provide such excellent outcomes, copious, nutritious pollen and a significant nectar surplus. A good Prickly Box flow will build strong hives ready for a leatherwood flow in January.

Flowers, what is the attraction? As humans, we often appreciate the beauty of flowers for their colour, form and fragrance. Flowers and their attraction to pollinators and bees in particular is much more diverse and complex. First seed bearing plants, angiosperms require pollination to occur. Flowers are the reproductive parts where seed production occurs and from which fruit and vegetables develop.

As part of this perfect design, flowers do many things to attract pollinators from what is visible to humans, to patterns in the ultraviolet spectrum, petal temperatures and textures and shapes. Research has shown that flowers and bees produce electric fields that also aid in attraction and bee flower memory (Robert et al 2013).

Experienced beekeepers know their flowers, especially those that deliver a honey crop. From years of observation they generally know the seasons and what is likely to flower, where and when. Urban flora is very diverse and contributes to "urban honey" a mix of whatever is available. Some beekeepers have even found that suburbs or at least their flora have a distinct flavour.

So what's here in Tasmania?

The 2004 Tasmanian Apiary Census captured valuable information about the flora used by bees and beekeepers in their apiary management. Managed honeybees make extensive use of all flora whether native or introduced. In Tasmania at least 93 species and plant associations were identified by beekeepers as being accessed by managed honeybees. However with urban beekeeping there are significantly more plants accessed that provide valuable nutrition.

Tasmania has a cool temperate maritime climate with many micro climates even providing subtropical conditions. From a simple drive by in any suburban street you will observe plants in flower from a global palette. Say in September or October, Ceanothus sp often called Blue Pacific, known as Californian lilac in its native California, is an understory shrub that is excellent bee forage producing

*A spring feast*

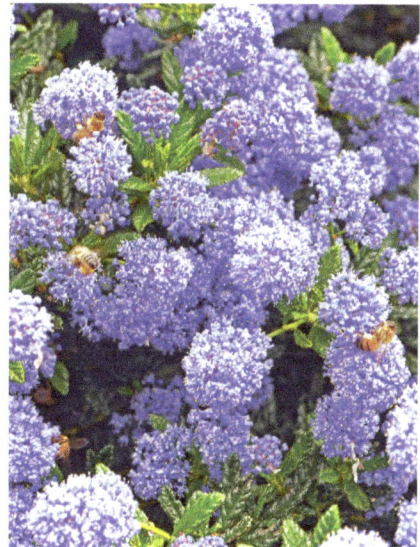

*Ceanothus is a bee magnet*

a unifloral honey to Eucalyptus caesia, Silver Princess, a most elegant weeping ornamental eucalypt from Western Australia attractive to honey bees also yielding excellent honey.

The list is long and includes street plantings, garden beds, herbs, vegetables, fruit trees, ornamentals and Australian and Tasmanian natives.

The following key findings provide some guide as to what is achieved by the apiary industry. This list is limited and many beekeepers have found other floral resources that yield unifloral honey often requiring careful observation of flowering times, sequence and specific management.

- At least 93 species and plant associations have been identified by beekeepers as being accessed by managed honey bees. There are many more when urban flora is considered.

- Leatherwood Eucryphia lucida is the most important floral element in the Tasmanian floral sequence, being accessed by at least 12,500 hives from at least 271 sites (2004 Industry Census). It is the latest regularly flowering species providing crucial winter stores.

- Blue Gum, Prickly Box and Manuka including other Tea Tree species (includes Leptospermum sp., Melaleuca sp. and Callistemon sp) are other native species that consistently produce sizeable commercial honey flows.

- The plants considered by beekeepers to be most important for pollen; Crack Willow, Gorse and Blackberry are all serious environmental weeds.

- Clover/Blackberry, "white honey", once the mainstay of the industry, is now threatened by changing agricultural practice and biological control.

*Silver Princess Eucalyptus caesia, a desert species that produces copius nectar*

*Leatherwood flower, simply elegant and an abundant producer of nectar and pollen.*

## Leatherwood

*Eucryphia lucida*

Leatherwood, Eucryphia lucida, an iconic Tasmanian nectar producer is a medium sized tree growing in cool temperate rainforest and mixed wet forest. Discovered as a honey producer by miners on Tasmania's West Coast in the early 1900's it is the main nectar producing plant in Tasmania. Its glossy green leaves form a compact crown and its abundant white four petal flowers are on show from early January until mid March. This varies with the season and geographical and altitudinal location. It occurs between latitudes 41°S to 43.5°S, from the Arthur River in the northwest to Catamaran in the south, It's best development is between 50m and up to 700m above sea level and requires at least 800mm annual rainfall.

# CHAPTER ELEVEN

Leatherwood honey is the mainstay of the apiary industry in Tasmania with annual production in the order of 1,000 tonnes and individual hive averages of 50-80kg per hive with an occasional bumper season. For Tasmanians the sites are remotely located in the southwest, west coast and northwest rainforest and mixed wet forest. The beginning of the season can often be a juggle between existing honey flows of Prickly Box Bursaria spinosa, Manuka, Leptospermum scoparium, meadow honey or pollination service. However, with intense nectar flows missing a day of good weather could mean losing 10kg per hive, a significant loss to larger commercial operations. Leatherwood apiary sites are sought after and are regulated by government land managers of production forest and parks and reserves. Small beekeepers do have the opportunity to place their hives on their Associations sites.

*Blue gum Eucalytpus globulus M Leech*

## Blue Gum

*Eucalyptus globulus*

The Blue Gum is the floral emblem of Tasmania and is one of the most widely planted Australian native plants. It mostly occurs as a medium to tall tree from 20m to 60m but occasionally much taller. However, in extreme coastal environments such as Cape Tourville on the East Coast it presents as a mallee like shrub.

The best sites for blue gum are in open woodlands where the trees are able to develop full crowns. The site should be sheltered but not damp and shaded, with good all weather access. Natural occurring blue gum dominated forest and woodlands occur in eastern and south eastern coastal regions, also forest edges and farm shelter plantings.

*Blue gum woodland M Leech*

*Blue Gum abundance M Leech*

Blue gum produces small volumes of honey and pollen regularly, with significant crops occurring every three to seven years. The tree demonstrates a strong sequential flowering beginning on the east side then moving to the west (Hoskinson 2010). The honey is light amber in colour with a heavy body and a distinctive flavour. The pollen is cream to white with a high crude protein content. Bees build well on blue gum sites (Leech 2012).

*Prickly Box is a primary coloniser of poorer sites*

## Prickly Box, Christmas Bush, Sweet Bursaria

*Bursaria spinosa*

Prickly box (Christmas bush, Blackthorn, Sweet Bursaria, Spiny Box, Whitethorn and more) is found in all eastern states including South Australia. It is a large shrub to small tree that is covered with thorns. It tends to occupy drier less fertile sites and is most prevalent throughout central north midlands, east coast and south-east Tasmania. It tends to flower from as early as late September in the southeast through to as late as early April in the northeast, typically available mid-December. The flowers are prolific is covered in dense pannicles of small white flowers to 1 cm in diameter, at the end of branches. The flowers have a pleasant sweet fragrance.

Honey flows in Prickly Box can be from nothing to extreme with reports of more

*A sweet fragrant delicate flower can produce abundant benefical pollen and nectar*

than 100kg per hive. This is very seasonally dependent and requires good rains and ground moisture prior to flowering. Attractive to many insects and birds it is an important primary colonising plant in drier environments.

Hives build very well on Prickly Box as the pollen is also very nutritious. This is a great pre leatherwood preparation flow that can also be very rewarding. Prickly Box honey is a very rapid crystalliser and the frames must be extracted as soon as 75% to 80% of the comb is capped. Failure to do this will result in comb that cannot be extracted. The filled frames can provide honey for the hive, but in the peak of the season this is not necessary and a waste of available drawn comb.

Prickly Box honey is a rapid crystallising honey with a fine texture that does set very hard. It has a most beautiful bouquet and considered a fine honey. For it to remain spreadable and still be considered a raw or cold extracted honey it must be creamed. Creaming as explained is a breaking of the crystal structure that prevents further crystallization providing a spreadable honey for a longer period.

*Manuka, Leptospermum scoparium, a Tasmanian native is widely distributed on poorly drained sites*

## Manuka

*Leptospermum scoparium*

Manuka honey now globally recognised for its well promoted medicinal benefits is a Tasmanian native. It is even thought that New Zealand Manuka originated from Tasmania and has become so well established that its highly valued honey has made naturally occurring sites very sought after. In Tasmania it too has become an increasingly valued honey. Once it was only considered a contaminant to leatherwood honey as it is dark and strongly flavoured and was only used to feed back to the bees. Now as people's appreciation of food has changed, its dark colour and strong flavour is sought after. Its highest value and biggest market is for bio-active Manuka honey that has been tested for its non-peroxide anti-microbial activity. Methyl glyoxyl has been isolated as one of the active ingredients, however, there remains much to discover about the relationship between the plant, its

*A Manuka thicket*

genetics, the environment it is growing in and the work the bee does in making it into honey.

Commercial beekeepers have formed the Tasmanian Active Honey Group to undertake further research into Manuka and other species including Leatherwood. They have also created a quality standard to aid marketing. There is potential to grow Manuka on poorer quality farmland that could provide significant returns from honey crops.

Other Australian Leptospermums have also been tested and can produce high levels of anti-microbial activity.

Manuka honey is dark amber, and is thixotropic, a jelly like state that becomes liquid when stirred or agitated. Thixotropic honey requires specially developed extracting equipment that pricks/agitates the cells. The honey crystallises slowly to a coarse granule. Manuka flowers annually and produces a honey crop in most years. Its pollen is a muddy white colour and produces average quantities of average to poor quality pollen.

The following description of Manuka honey is adapted from (www.honeytraveller. com 2011)

Manuka honey is distinctively flavoured with thixotropic 'gel-like' properties and becoming less solid when stirred or shaken. It is slow to granulate and forms coarse crystals and is often creamed. Its colour is dark cream to dark amber. It is aromatic with damp earth and heather notes and a cool menthol (or eucalyptus) taste and rich flavour of mineral, barley sugar and herbs. It is medium sweet with a slightly bitter aftertaste. In its creamed form the honey has a cool, smooth feel in the mouth. It is particularly good as comb honey.

Plantations: There is significant potential to plant Manuka and other known active Leptospermum sp on farms as riparian zone plantings, outside components of taller windbreaks, low shelter or as larger monocultures found naturally. Work is being done in New Zealand to find out how to produce active manuka honey more consistently with higher yields from Manuka plantations. www.manukafarmingnz. co.nz

*Open grown Blue Gum has a large crwon area with progressive, abundant flowering*

## Tasmanian Eucalypts

The Tasmanian eucalyptus species vary in their production of pollen and nectar. Tasmanian Blue Gum is perhaps the most reliable producer yielding smaller amounts most years and a major crop year every 7 to 12 years.

The floral calendar for native and non-native flora provides information gained from an apiary industry census 2004. It is a regionalised table with information on flowering frequency, duration and relative value of pollen to the bees and average yields of honey.

There is a wide variety of native species that are beneficial to bees but may not yield a surplus. Many of the 'pea-flower' leguminous understory species provide valuable, nutritious pollen as do some of the acacia species.

| Common Name | Botanical Name | Distribution | Flowering Interval yr | Floral Calendar | Pollen Value | Honey Yield |
|---|---|---|---|---|---|---|
| Black peppermint | E. amygdalina | Widespread | 8-15 | ONDJ | Poor to average | 0-35kg |
| White top | E. delegatensis | Higher altitudes | 1-20 | ONDJFMA | Average 23% cp | 0-38kg |
| Tasmanian Blue Gum | E. globulus | East, southeast | 1-12 | JASONDJ | High 29% cp | 5-60kg |
| Stingybark | E. obliqua | widespread | 7-14 | FMAMJ | Average 24% cp | 3-38kg |
| Black Gum | E. ovata | widespread | 1-4 | JASOND | Poor 18% cp | 8-40kg |
| Ironbark | E. sieberi | Mid east | 6 | ASON | Poor | 0-30kg |
| White Gum | E. viminalis | widespread | 2-10 | NDJFMA | Low to High | 0-30kg |

Other native species that yield a surplus.

There are a few commercial beekeepers who specialise in identifying unifloral honey from other than the main producing species.

Mountain Pinkberry, *Leptecophylla juniperina* is a winter flowering species that requires specific colony management to produce a surplus. With poor flying conditions, lower temperatures and a low sun angle, hive location and orientation is critical. Beekeepers that persevere to discover the requirements of specific species can produce exquisite, unique unifloral honey.

*There is a large number of non-native species from herbs, shrubs and trees that provide beneficial forage for bees and other pollinators.*

Non-natives

There is a large number of non-native species from herbs, shrubs and trees that provide beneficial forage for bees and other pollinators. Often they do not occur in large enough numbers to produce a surplus, but are known honey producers in their homeland. Some agricultural crops and horticultural orchards and plantings can produce significant honey flows.

In Tasmania there are some widespread environmental weeds that provide beneficial bee forage, especially abundant high value pollen at important times of the year. In some cases such as crack willow, Salix fragilis are being removed from the environment by large scale programs as this species has infested many river environments. It has been one of the prime sources of early season highly valued abundant pollen but is now much less available. European gorse, Ulex europea is a widespread weed in agricultural and forest environments that also provides high quality pollen through winter and early spring. Blackberry, Rubus fruticosus famous as a producer of high quality honey also provides high quality pollen.

## Urban environments

Public spaces; parks, public gardens, and your garden are all places of potential bee-forage. Bees are not limited to your backyard will fly many kilometres for favourable food. However, in most city and urban environments they tend to have an abundance of forage and usually fly less than 500m. Therefore, in looking at available resources for your bees you could take a neighbourhood approach and you will be surprised how much variety there is. Most urban beekeepers will stay focussed on the urban environment rather than taking on the tasks of chasing honey flows or migratory beekeeping practiced by the overwhelming majority of commercial beekeepers.

Planting plants for pollen and nectar supply.

*Crack Willow, Salix fragilis, the subject of a removal program, provides abundant early pollen, ideal for spring build.*

*Blackberry, a common environmental weed was a mainstay of the industry*

*Nature strip plants provide important urban bee forage*

*A bee friendly sidewalk planting*

*Larger clumps provide a better foraging outcome*

We can all make a difference, even a pot plant on a balcony can be beneficial. There are numerous lists and guides for bee friendly plants. The book Bee Friendly: A planting guide for honey bees and Australian native pollinators (Leech 2012) is a very useful starting point. As a beekeeper you should be aware of what plants provide good nutrition. The table below provides a brief list of beneficial plants for urban environments with an indication of their nectar and pollen value to honeybees.

More comprehensive information can be found in Bee Friendly (Leech 2012), the Xerces Society and ask your local nursery

*Bee Friendly, M Leech 2012*

| Plant (Common Name) | Pollen/Nectar |
| --- | --- |
| **Home orchard** | |
| Almonds | P |
| Plums | n P |
| Apple | n P |
| Cherry | n P |
| Lemons (citrus) | N P |
| **Herbs** | |
| Basil | n p |
| Borage | N P |
| Coriander | N |
| Lemon balm | N p |
| Mint | N P |
| Thyme | N P |
| Rosemary | N |
| Marjoram | N |
| Oregano | N |
| Sage | N |
| **Vegetables** | |
| Beans | np |
| Squash family (pumpkin, zucchini) | N P |
| **Berries and Currants** | |
| Black and Red Currants | N p |
| Blueberry | N |
| Raspberry | N P |

| Plant (Common Name) | Pollen/Nectar |
| --- | --- |
| **Ornamentals** | |
| Aster | |
| Catmint | N p |
| Crocus | P |
| Echium candicans | N p |
| Escalonia | n p |
| Ceanothus sp | n p |
| Cornflower | n p |
| Hebe | N P |
| Iceland Poppy | N P |
| Lavender | N p |
| Sunflower | N p |
| Zinnia | n p |
| **Trees** | |
| Linden/Lime | N |
| Sweet Chestnut | N P |
| Tulip tree Liriodendron tulipifera | N p |
| **Native Plants** | N p |
| Mallee eucalypts (small) | N P |
| Eucalyptus caesia | N P |
| Corymbia ficifolia | N |
| E. leucoxylon rosea | N P |
| Banksias (ericifolia a must) | n p |
| Callistemons | n p |
| Grevilleas | n p |
| Hakea | n p (can be N P) |
| Leptospermum sp | n P |
| Native Peas | 0-30kg |

and talk to fellow beekeepers as they know what their bees are working.

Planting bee forage for all seasons contributes significantly to the productivity of the garden, and the cumulative effect of increased quality bee forage will ultimately help produce healthier bee populations of managed honey bees, native bees and other pollinators.

*Western Australian Red Flowering Gum, Corymbia ficifolia one of the most widely planted trees in the world is a bee magnet*

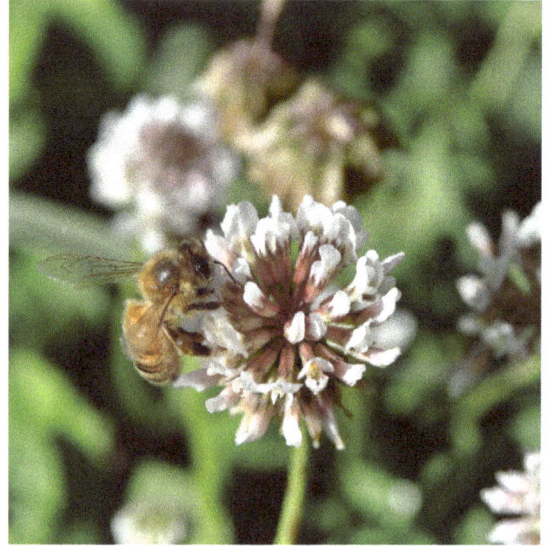

*Clover provides a continuous flowering*

*Oregano in a garden pot*

As an urban beekeeper it is good to be involved to some degree in gardening as you can observe the interaction between the plants you are growing and their benefit to honeybees. You can also be an advocate for bee friendly planting, the more that sing this song, the better the planted outcome across the landscape.

A bee garden can be as limited as a single pot plant or as extensive as an entire garden, complete with large trees, ornamentals, shrubs and vegetables. Vegetable gardens and home orchard yields can be improved by planting bee-friendly plants between rows or at the ends of beds (Barrette 2010).

Bee garden design criteria

Many authors say that the most important design criteria in a bee friendly garden is floral variety and continuous flowering.

The following is taken directly from Bee

Friendly (Leech 2012).

*Bees like a varied diet, so plant many flowering plants that are beneficial to bees, native plants and 'heirloom' or open pollinated varieties, avoid modern hybrids and 'pollen-free' plants. It is important to have at least four different species flowering at any one time and to have continual flowering throughout the year. In cold climate regions, abundant winter flowering can cause problems for the colony (Somerville 2002).*

Bee plantings should be in multiples or clumps, in layers from the ground up to trees (Barrette 2010). Its preferable to plant larger swathes, wide borders and beds filled with a variety of flowering plants in flower throughout the year.

General Gardening Advice for Attracting Bees and Other Pollinators

1. Don't use pesticides. Most pesticides are not selective. You are killing off the beneficial insects along with the pests. If you must use a pesticide, start with the least toxic one and follow the label instructions to the letter.

2. Use local native plants for native bees. Research suggests native plants are four times more attractive to native bees than exotic flowers. They are also usually well adapted to your growing conditions and can thrive with minimum attention. Exotic plants, most vegetables, herbs and fruit trees provide nutritious pollen and nectar for honeybees. In gardens, heirloom varieties of herbs and perennials, single flower varieties can also provide good foraging.

3. Chose several colours of flowers. Bees have good colour vision to help them find flowers and the nectar and pollen they offer. Flower colours that particularly attract bees are blue, purple, violet, white, and yellow.

4. Plant flowers in clumps. Flowers clustered into clumps of one species will attract more pollinators than individual plants scattered through the habitat patch. Where space allows, make the clumps one metre or more in diameter.

5. Include flowers of different shapes. Open or cupped shaped flowers provide the easiest access and shorter floral tubes are important for honey bees. Other pollinators including native bees, butterflies and birds benefit from differing flower shapes.

6. Have a diversity of plants flowering all season. A varied diet is essential for the well being of honey bees and other pollinators.

7. Plant where bees will visit. Bees favour sunny spots over shade and need some shelter from strong winds.

8. Provide accessible water Bees need access to water. Provide easy access either through wet sand or pebbles, don't drown the bees.

*Adapted from the Pollinator Conservation Program, Xerces Society Shepherd (2004).*

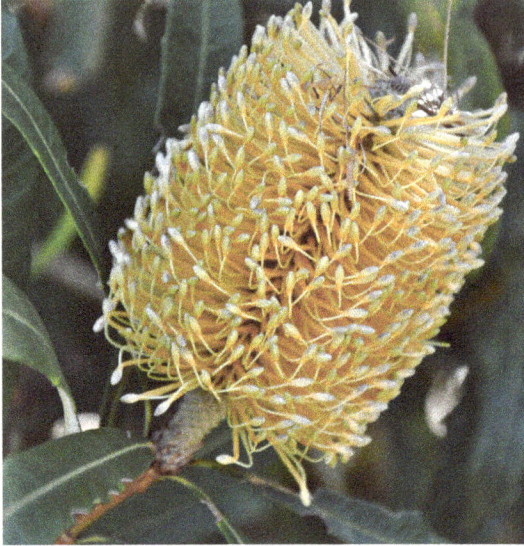

*Coast Banksia, Banksia integrifolia, a prolific autumn nectar and pollen producer.*

*Happy Wanderer, Hardenbergia*

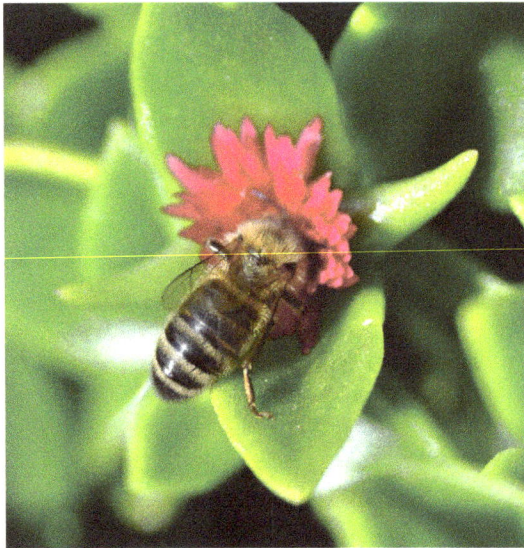

*A form of miniature pigface that is a bee magnet*

*Cherries pollinated commercially by managed honeybees*

*Lavender, always with bees.*

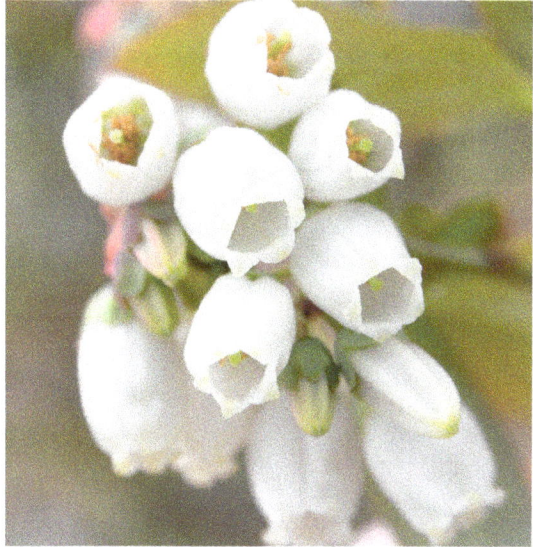

*Blueberry, honeybee pollination produces better fruit*

*Japonica, produces the first flowers still in winter*

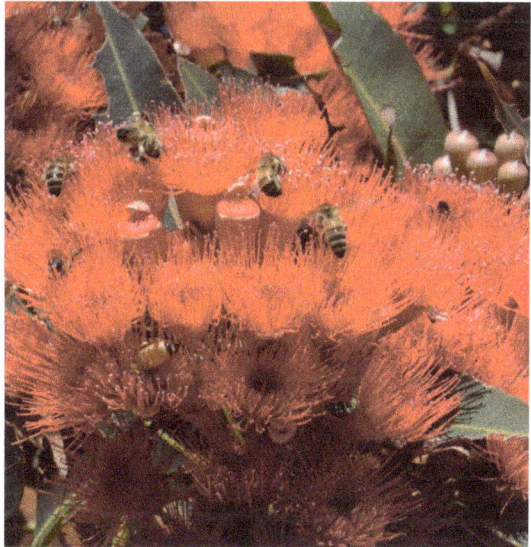

*South Australian Blue Gum, Eucalyptus leucaoxylon, a late summer abundant nectar source*

*A purpose bee planting. Des Cannon*

*Coral Gum, Eucalyptus torquata, bee planting in the Negev Desert, Israel. Dan Eisikovich*

Late winter and early spring flowering is important to provide pollen and nectar as the colonies begin to build their numbers and require an increased food supply, having depleted their winter stores. Similarly, a summer gap can occur and it is important to maintain the food supply.

Good nutrition, a varied diet and constant availability through late winter, early spring over summer and into autumn is the goal that we are aiming for across the landscape. A high quality diet is one of the primary safeguards against disease and recovery. Keep a diary of what is flowering where and when, overtime this will build to become an invaluable resource. Think of creative ways to enhance bee forage, encourage local governments, schools, farmers and other land managers to grow plants that will be beneficial to honey bees. You are the bees best advocate, the time to plant is now, we can all make a difference!

With one hive it all seems like a free kick, the wonderful workers do it all, well most of the work and provide you with an amazing surplus of honey. Remember, most people get into beekeeping because of the bees and get out because of the honey.

Even one lonely hive can produce relatively large volumes of honey when there is an abundant nectar flow or series of flows or you move your hive to catch honey flows and maximize honey production. The industry average in Tasmania is approximately 50-80kg per hive and well managed amateur hives can match this, but are usually less. In peak seasons these volumes can be exceeded. If you were used to your gentile 30kg plus from your stationary garden hive, then 80kg of honey is quite confronting and then multiplied by three or four there is some serious work to be done. Your attitude and approach all depends on your reasons for keeping bees.

But for now it is time to roll up your sleeves. Nectar flows in Tasmania usually don't commence until late spring or early summer. There are some specific plants that flower in winter such as Eastern or Pink Mountain Berry but harvesting a honey flow requires very specific management. Early spring blooms where in abundance such as orchards rarely produce a surplus, but for most amateur beekeepers your main honey producing activity will take place from late spring to early summer.

Having built your hive up for the honey flows with enough supers for storage as described in chapter ten Seasonal Management, it is time to harvest. Harvest of a few frames can be undertaken earlier in the season, but over robbing early is not

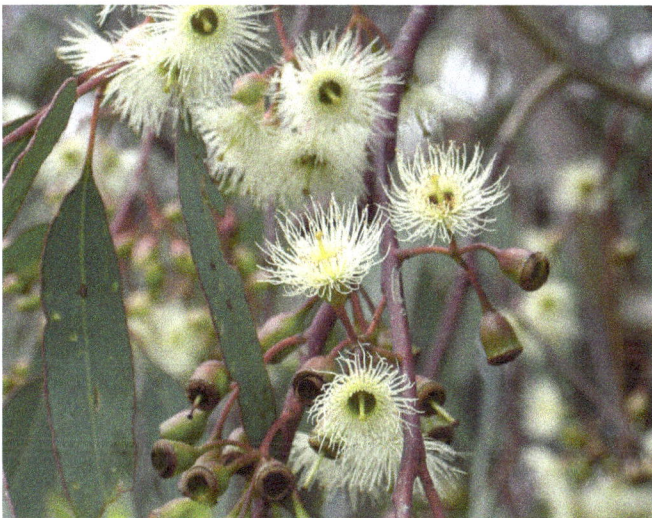

*Red Ironbark, Eucalyptus tricarpa, is an abundant nectar producer but lacks pollen*

*A single hive prepared for Leatherwood*

recommended as the weather can deteriorate for weeks at a time, then all stores are needed for the colony.

November is the time for early Manuka, Blue gum, clover and blackberry to begin flowering and there are some strong urban contributors such as Eucalyptus tricarpa/sideroxylon Red ironbark, Echium candicans, Pride of Madeira and Ceanothus, Blue Pacific.

Harvesting / Robbing

Harvesting your urban honey can also be undertaken at the end of the season as it is unlikely that a unifloral honey will be produced, given the different mix of flowering over time. Flows from single or predominate species producing unifloral honey such as clover, blackberry, blue gum, manuka, prickly box and leatherwood are harvested at the end of the flow or on a regular basis, even weekly on heavy flows.

Some require immediate extraction, canola and prickly box crystallize very rapidly and if left on for too long will crystallize in the comb, effectively un-extractable, only available to the bees. This becomes very important for those chasing honey flows as bees are usually taken from prickly box to leatherwood, crystallized honey in cells takes up space and is lost production from leatherwood or the next nectar flow.

Harvesting of your honey crop can be undertaken during a honey flow when there is an average of at least seventy five percent of capped comb, any less and the

*A frame with minimum capped cells for honey extraction*

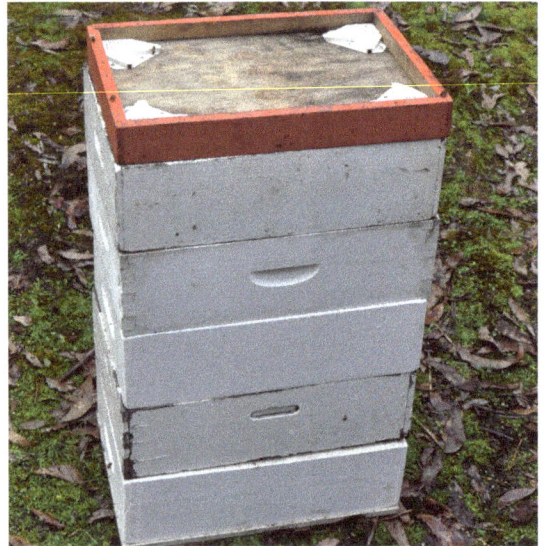

*A clearer board used prior to robbing*

honey will not have been cured enough and the moisture content will be too high for storage (greater than 18% moisture content), resulting in fermentation.

Pick your weather. The process should be undertaken on a warm sunny morning when the field bees are busy foraging. Approach your hive from the side or the back, gently puff some smoke into the entrance, lift the lid slightly and gently puff again, lift the mat and check the frames in the top honey super.

Clearing bees can be done in a number of ways, in urban environments gentle is best and the use of a clearer board placed underneath the honey supers allows the bees to travel down but the one way entrances prevent them from returning.

This may take a few hours, but many beekeepers like to allow overnight. This is not a practical solution if your bees are far away. After most bees are removed clearer boards are placed on stacks of supers being transported back to the honey house for extraction,

Another method suited to small operators is to remove a frame at a time, brushing the bees off with a bee brush and placing the cleared frame in another super in an enclosed container. Motorised blowers are also used to remove bees.

This is slow and does not work if you have more than one or two hives.

Wacker frame: A steel frame with rubber padding. In areas removed from urban environments it is more efficient to remove all the bees from the supers as quickly as possible.

*A soft bee brush used mainly by amateur beekeepers to gently remove bees*

*A frame used to dislodge bees prior to robbing*

This can be achieved by using a strong metal or wooden frame with enough rubber or foam to soften the blow. You jolt the super onto the padded frame resulting in the bees falling to the ground in front of the hive. This is a very efficient method and the bees are back in their hive in a short time.

Extracting

Plan the extracting process well as you can end up with a sticky mess. What will

*Plan the extracting process well as you can end up with a sticky mess.*

you cover the floor with or can it be washed, where will you place the supers and extracted frames, where will you store honey, do have you extra containers, work through this exercise well to avoid the avoidable!

Now that the frames have arrived at your extracting location the honey will soon start to flow. But first you need to uncap the cells within each frame before placing them in the extractor.

Uncapping: Simple uncapping is achieved with either a water, steam heated or electrically heated uncapping knife.

If you are using a water heated one, it is more efficient to have two knives so one is always in the hot water ready to reuse. You also need a container to catch the wax cappings and honey, preferably one with a strainer and a honey-gate. A wooden brace with a screw point sticking through it vertically as a pivot provides a simple means of securing each frame as you hold the frame with one hand and move the uncapping knife along the frame with the other. Tilt the frame and the hot knife will slide over the edges of the frame, wax and honey will fall into the container. Once you have done both sides the frame can be placed in the extractor.

Commercial uncapping is done with automated or semi-

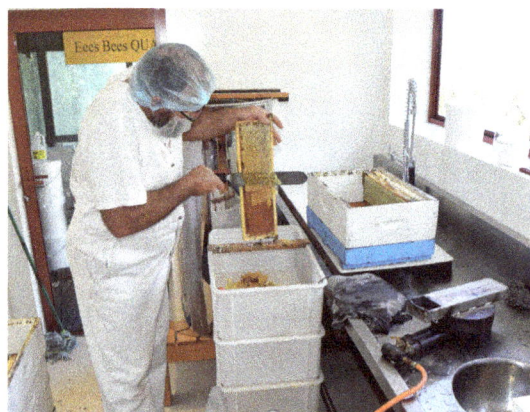

*A water heated uncapping knife*

automated electrically driven uncappers, these come in different configurations and fit within a system of efficient wax and honey flow within the honey house.

Extracting: there are a number of different extractors suited to the smaller beekeeper from manually cranked versions to small radial variable speed electric extractors. Small manual extractors can be hired on a daily basis and your local beekeepers association may have one to borrow.

The extractor must be evenly loaded before spinning commences. It is very important to start out slowly as fast initial speed will end in broken comb making extraction very difficult.

You will see honey spin out onto the sides of the extractor. If using a tangential extractor only do half of the first side before swapping them around. Gradually increase the speed until all honey is removed from the second side. Then return to the first side and complete extraction.

*A small manual tangential extractor*

With radial extractors frames are oriented out from the centre, like spokes in a wheel and honey is extracted from both sides simultaneously. This style scaled up is most common in commercial honey houses.

Once honey is extracted it is initially filtered and stored.

# CHAPTER TWELVE

*A 12 frame radial extractor, efficient for a 30 hive apairy*

Leaving it to settle overnight in a warm room will allow air bubbles and any remaining wax to rise to the surface where it can be scraped off providing a clear final product.  Convenient cost effective storage for smaller beekeepers are 20l plastic food grade sealed lidded buckets.

Note these must be filled, sealed with duct tape to make airtight otherwise the moisture content of the honey tends to increase over time and can lead to fermentation.

Commerical producers and packers typically use food grade stainless steel tanks for storage.

Honey gates are very useful as they enable you to bottle honey directly from the storage container with minimal mess.

*20l food grade plastic buckets sealed with tape to prevent moisture entering*

*Stainless steel storage and bottling plant*

*Basic filtering using 'shower screen' mesh fabric*

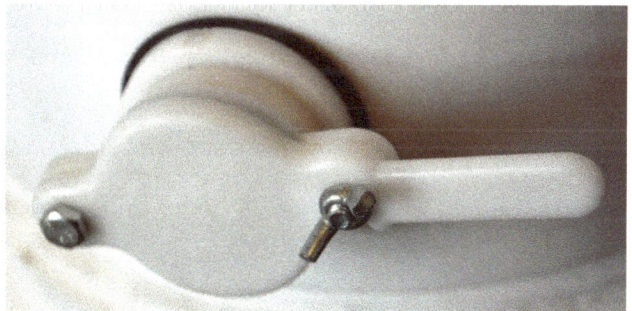

*Plastic honey gates are commonly used*

## Storage

Honey can be stored for many years in clean airtight containers. Ensure your containers, glass, plastic or metal are food grade and have been sterilised. Bottling after allowing the honey to settle but while still in a "runny" state minimises double handling and gives you product ready to sell. It does mean that you have had to purchase all your containers at once. This may be cost prohibitive and you may have to stagger your bottling. When filling containers for retail sale it is important to hold the jar or tub as close to the tank as possible, slightly tilted so honey runs down the side, minimising air bubbles. Air bubbles can spoil honey appearance and induce crystal formation (Ayton 1991).

*The presentation of your honey is a key to successful sales*

Marketing & Presentation

The presentation of your honey is a key to successful sales, especially as the food market has become more informed and there are more boutique outlets selling

*Well presented product helps it stand out*

*Honey Tasmania Shop is a specialty honey shop that delivers great brand management, product display and service*

quality, well presented product.

You should do your marketing, understand what market you are selling into and what presentation is most appealing. You only have to look at wine labels to get some understanding of how companies try to gain a marketing advantage through their label. An attractive label can tell a story and give you a further opportunity to talk about your product at markets. It is also the first impression that commercial buyers and retail customers will see. Remember in most retail situations they cannot taste your honey and you are not there to explain its unique qualities.

Food Label: by law you are required to have a food label, a standard honey label

| NUTRITION INFORMATION | | |
|---|---|---|
| 370g | | |
| Servings per pack 53 | | |
| Serving size 7g | | |
| | Ave Quantity Per serve | Ave Quantity per 100g |
| Energy | 98kj | 1400kj |
| Protein | 0.0g | 0.2g |
| Fat,total | 0.0g | 0.0g |
| -saturated | 0.0g | 0.0g |
| Carbohydrate | 5.8g | 82.1g |
| -sugars | 5.8g | 82.1g |
| Sodium | 1mg | 14mg |
| Packed by | | |
| Adresss | | |

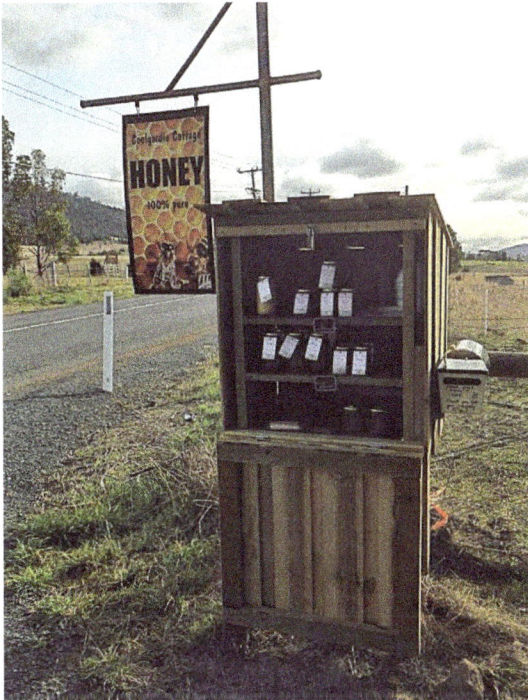

*A road side stall of excellence. First impressions may attract customers to your stall*

*A food label is a legal requirement*

is available that provides all the required nutritional information. You will have to provide the serve size and number of serves in your container.

Food Safety: When selling to the public it is advisable to have a food safety plan. Most commercial producers will have a formal HACCP Plan. HACCP is an internationally recognised method of identifying and managing risk and, when central to an active food safety program, can provide vendors, the public and government sector bodies with a high degree of comfort that food safety is being taken seriously and is well managed. (www.haccp.com.au 2016).

Selling

How much should I sell my product for? Do not be guided by supermarket prices. Visit boutique food stores and remember you are looking at retail prices and will most likely get fifty to seventy percent of what you see as a wholesale boutique food price. If you're going to sell at markets, famers markets or regular weekend markets go and check them out, see what's selling, how it's presented and for what price. Talk to the beekeepers, that's always a good start.

Remember, you were once a buyer of honey and maybe wanted to find out more, the world is very interested in the bee story, get engaged, this is your opportunity to be an ambassador for bees and their wellbeing.

Honey Styles

Honey is produced in a variety of styles. Most small producers cold extract and leave their honey to crystallise in the jar, a normal process and one that helps sell the 'raw' attributes.

- Liquid 'runny' honey: Most Tasmanian honey will crystallise. Some of the eucalypt honey varieties stay liquid for a significant time. Honey should not be heated above 65°C, if you want to maintain its natural aroma, volatiles and heat sensitive enzymes it should not be heated above 35°C, the brood temperature maintained by the bees.

- Crystallised or candied honey: the easiest to produce, just package and let it be! Some honey will crystallise very rapidly such as prickly box, it is advisable to pack it soon after extraction.

- Creamed honey: This can be achieved in a number of ways, either cutting crystallised honey into small sections and whipping it in an industrial food mixer, or seeding liquid honey with fine grained crystallised honey.

- Comb honey: is capped honey in the wax comb, honey at its purest. This does not require any uncapping or extracting equipment, but may need particular management and the use of unwired thinner foundation. Handling the product requires great care as jarring or dropping will easily crack the fragile wax cappings and honey will leak. (Wellington Apiaries).

• Honeycomb cannot be shipped to the Australian mainland unless treated, due to the presence of Braula fly in Tasmanian hives. It must first be frozen for at least 48 hours to brake the braula fly lifecycle. A vendor declaration must then be completed prior to shipment.

• Cut comb: some beekeepers simply de-wire their comb and cut sections to be presented in small packages. Once cut the edges should be drained overnight providing a 'dry' surface.

*Well packaged honey comb ready for dispatch. Wellington Apiaries*

- Sectioned comb: Is provided in small wooden, plastic or metal frames, and produced in half supers. It requires more specific management to ensure the bees draw the comb and fill it with honey. This may be better achieved using first swarms. Different techniques can be used to encourage established bees to work the sections of comb, requiring a strong colony and a heavy honey flow. Queen excluders should not be used initially and the hive reduced to two ideal or one full depth super of brood with the section-super over the top. When this is half full a second section-super is place under it. To minimise swarming add empty section-supers on top of the broodnest. Remove fully capped section-supers as soon as possible as bees walking across the cappings mark them making them less presentable.

- Comb in honey: simply place cut comb in a jar of honey, some people love it.

### Valuing adding to honey

Value adding to honey beyond presentation and honey itself is often related to baked goods of some form. Honey and bees wax provide other outcomes including cosmetic application and is mentioned under wax.

Traditional honey baked goods is described in the chapter Honey. Cooking with honey has been practiced by many different cultures over thousands of years. It was mentioned in Ancient Egyptian writings from the reign of Ramesses III (1200-1085BC) as an ingredient in cakes and often used by the Roman epicure Apicius in his cooking book from the first century AD. (Crane 1999). Today the use of honey is very visible in sweet treats from Mediterranean and Middle Eastern countries, the many nougat styles from France, Italy and Spain, not to forget honingkoek (honey cake) from Holland, ginger bread, lebkuchen and marzipan from Germany.

*German Lebkuchen a traditional use of honey in some contemporary figures.*

Jewish custom has many festive sweets using honey and Russian honey cake is amazing. It is also used in classic Asian cooking from India to China. This is not to tease you but why not try using some of the honey you produce that is less favoured by the market and turn it into something exotic and irresistible.

Fermented drinks

- Mead is an ancient drink with first evidence of its use dating to ancient Chinese pottery vessels of 6500-7000 BC, McGovern et al (2004) and European use dating back to 2000 BC, Jazz (2009). It has been continually used as a simple fermented honey product and has many variations in cultures throughout Europe, Asia and Africa. It has often been associated with monasteries that kept up the tradition of mead making as a by-product of beekeeping. There are many varieties of mead with flavours depending on the honey source and what fruit or herbs have been added. There is an increasing trend to appreciating mead with a number of producers of high quality mead in Tasmania.

*Honey based alcoholic drinks are becoming increasingly popular*

- Beer has also been made with honey at least as an ingredient since ancient times. Honey beer in its simplest form is made in Africa in a few days, simply from honey water and dry yeast. Similar to its Ancient Egyptian history where woman or beer wives made the brew, in Africa, beer ladies tend to be the brewers. Its popularity is increasing as boutique breweries abound. Here in Tasmania Taverners Brewery have been pioneers and produce award winning honey beers.

- Other honey beverages such as liqueurs have been produced for centuries and include some famous brands; Baerenjaeger a German liqueur produced since the 15th Century, Krupnik a 16th century honey liqueur from Poland and in Drambuie in Scotland. Drambuie, originated in the 18th Century as Bonnie Prince Charlies secret recipe, given as a thank-you to John McKinnon chief of the McKinnon clan, more recently commercialised, and Glayva from Scotland. A number of unique Tasmania honey liqueurs have emerged along with Tasmania's new boutique distillery industry.

Other hive products

There are a number of other hive products including pollen, propolis, royal jelly, wax and the bees themselves. As a small beekeeper the most important product other than honey will be wax and its value adding opportunities.

Pollen: is essential for brood rearing and while it has a market as a health product for human consumption and as bee food, small apiarists may find pollen collection more trouble than its worth. Protein rich products lose their nutritive value rapidly at room temperature, so regular collection and correct storage is essential to maintaining its value. Some commercial apiaries specialise in the production of pollen and pollen patties for feeding bees, others producing it for the health supplement market. Perhaps a more important reason for collection is to check for possible pesticide contamination and also to identify the floral resources that the bees are accessing.

*The chemical composition of wax is complex with more than 180 compounds identified.*

A simple trap is used at the hive entrance, bees are forced to enter through a grid that causes the pollen to brush off and fall into a collecting drawer. Ideally pollen collecting should only be done when there is pollen abundance and only for short periods allowing bees to replenish their own stores. Freshly collected pollen can be frozen and fed back to the bees once defrosted. If you dry pollen either in the sun or artificially heated it can be stored at room temperature but it will have to be mixed with sugar to be attractive to bees.

Wax

Beeswax is collected as you go about your honey business. As you prepare frames for extraction and uncap cells, you produce a wax and honey mix. This is the main source of clear wax and it is initially separated through a course strainer. Other wax can be collected through removal of burr comb at each hive inspection and stored for rendering and sale, or production of wax based products.

*Bees wax has been used in candle making since the Middle Ages*

The chemical composition of wax is complex with more than 180 compounds identified. An approximate chemical formula for beeswax is C15H31COOC30H61 consisting of hydrocarbons, straight-chain monohydric alcohols, acids, hydroxyl acids, oils and many other substances. Bees wax has a melting point of 61-66°C and a flash point of 254-274°C depending on its purity. Its specific gravity is less than 1 so it floats in water. Wax is insoluble in water, slightly soluble in ethanol (alcohol) and soluble in benzene. As archaeologists have discovered beeswax is stable for thousands of years.

Melting small volumes of wax can be done in a cloth bag in hot water, you need to add some old honey to the water to ensure an acidic solution and a clean wax outcome otherwise you create a saponified layer. You can also make a simple solar wax melter. Plans for a very inexpensive and simple wax melter are available from the ACT Beekeepers Association or more sophisticated wax melters are available from beekeeping equipment suppliers. However you decide to do it, your melted wax can be poured into suitable containers for producing larger blocks of wax that can be later melted and moulded into the shape or incorporated into the products you are making.

*Liquid beeswax is highly flammable and should never be heated over a naked flame, only use a double boiler or water jacket to safely melt wax.*

Caution

Liquid beeswax is highly flammable and should never be heated over a naked

*An innovative candle in a tin*

*Bees wax can be a safe base for cosmetics*

flame, only use a double boiler or water jacket to safely melt wax.

Beeswax has many uses and has provided a traditional base for sealing as leather conditioning, furniture polishes, as a modelling material for lost wax casting, in candles, cosmetics and mostly for wax foundation for the bees.

Beeswax Product

- Wax for foundation is rolled and impressed as a starter for the bees to build their wax cells, to draw new comb.

- Candles from beeswax have traditionally been used in the Catholic and Orthodox churches and in their infinite variety can be a useful sideline product for small beekeepers. Beeswax burns hotter than candle wax and may cause glass containers to crack. It does not cause soot, they do not drip, burn slower and are said to purify the air. There are many useful internet sites about beeswax candle making. Cosmetics: beeswax is used throughout the cosmetics industry to produce lipsticks, lip balms, skin nourishing products, soaps and shampoos. It is possible to make some simple cosmetic products on a small scale, again the internet is a rich source of information or recipes can be found in many books such as (Flottum 2011.)

*Propolis may have an antimicrobial and antifungal affect within the hive helping to prevent disease.*

Propolis

Propolis, the sticky dark stuff that makes it hard to separate supers and frames, bee glue is collected by bees from plant exudates that are in a semi liquid state; from wounds, buds and bark of shrubs and trees. It is collected by foragers on their pollen collecting apparatus and removed by house bees inside the hive and mixed with one third bees wax and an unknown substance, (Sammataro 2011.) Bees produce this sticky resinous substance to seal cracks in the hive to prevent heat loss, mend comb, glue frames and fill gaps. Caucasian bees are well known for their propolis collecting ability but excessive use in the hive makes hive management sticky and difficult. Propolis is a product in demand by the health supplements industry and can be simply collected by using a propolis trap, a slotted sheet that's sits on the top bars of your top honey super taking advantage of the gap filling imperative of workers. When full the sheet can be frozen and the propolis knocked out and stored in a glass jar.

# CHAPTER TWELVE

*Royal Jelly, the food of queen larvae has reported health benefits*

Propolis may have an antimicrobial and antifungal affect within the hive helping to prevent disease. This will vary with location and what trees and shrubs the bees are collecting it from.

## Royal Jelly

Royal jelly is the white gelatinous substance that larvae and the queen bee are fed by young worker/nurse bees. It is the product of the hypopharyngeal and mandibular glands. It consists of water, protein, lipids (fat), carbohydrates vitamins and other compounds. Collection of royal jelly is from queen cells when the larvae is three days old, the time when the most jelly is present. It can be collected by the amateur beekeeper by raising queen cells to collect it. Its commercial production is a specialised form of beekeeping, but a well-managed strong hive should be able to produce 500g in a five month season. Tasmanian experience using a professional royal jelly apiarist produced 500g per day from 500 hives, it was noted that the hives need a recovery period, (Norris 2015.) Royal jelly has numerous precious therapeutic properties used from ancient times until today. The product has little clinical evidence to support the therapeutic claims but remains a very popular health supplement. (Pavel et al 2011).

## Venom

Bee venom has a limited medical market used in desensitisation. Venom is collected from individual bees via an electric shock and a sting response. Collection is onto a glass plate where it dries and can be scraped off for storage. Great care must be taken in handling it, plastic gloves, eye and breathing protection. Bee venom treatment has been reported to cure forms of arthritis and skin disorders. Any application must follow clinical advice and be aware of possible anaphylactic reactions.

It takes a million bees to produce 1 gram of venom but it is said that twenty hives can produce one gram in two hours (Dotimas and Hider 1987).

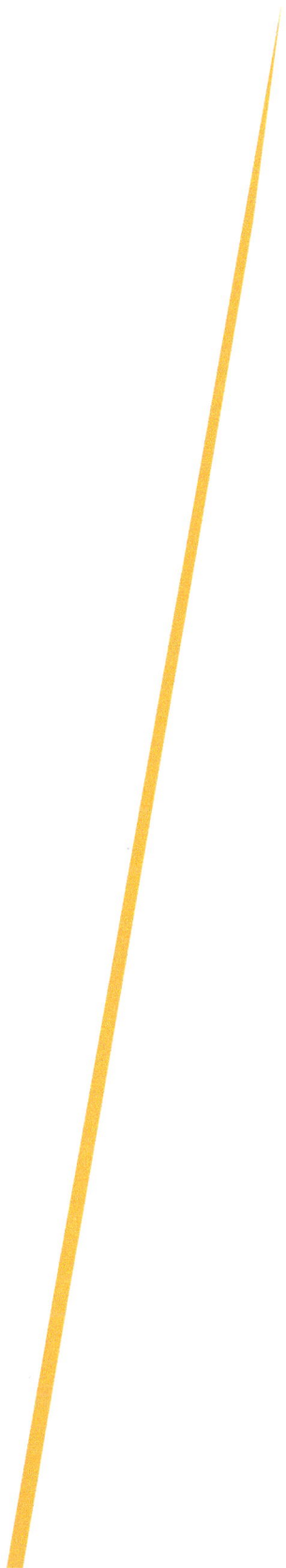

In the introduction it was said that most people start beekeeping for the bees and stop because of the honey. That relates more to an unrealistic expectation of the amount of work you need to do to extract, store, bottle, market and sell your honey. It all relates to scale, because as a single or two hive urban or migratory beekeeper, it takes a minimum commitment of time even without a big honey flow. It is most likely that your costs of production, maintenance and replacement outweigh your returns from beekeeping, but perhaps this overlooks the fun, pleasure, intellectual reward, generosity and stewardship gained from having bees.

A good place to start in a discussion of honey is the accepted international standard definition from the Codex Alimentarius, the international food standard commission of the United Nations Food and Agriculture Organisation FAO.

Honey is the natural sweet substance produced by Apis mellifera bees from the nectar of plants or from secretions of living parts of plants or excretions of plant sucking insects on the living parts of plants, which the bees collect, transform by combining with specific substances of their own, deposit, dehydrate, store and leave in the honey comb to ripen and mature (Codex standard:1987).

There is an increase in understanding and appreciation of varietal or 'unifloral' honey that could be expressed as honey the new wine i.e., honey that is from a

*Leatherwood honey, it doesn't get any fresher*

*Honey gatherer from cave painting near Valencia, Spain estimated 8000 years old.*

predominant floral source with unique characteristics and provenance. If you are a beekeeper intent on exploring honey production, not as a provider of commercial pollination services then exploring and experimenting with the opportunities to provide a number of unique honey varieties in a season will captivate you. The market for different colours, flavours and consistency of honey has increased, and honey that was once considered bee food or a contaminant to a traditionally preferred honey is becoming more appreciated. A well-known example is Manuka, Leptospermum scoparium, which is highly valued and sort after for its health benefits but also as a table honey for its thick consistency, dark colour and rich taste.

The development of honey from the flower to the jar is a fascinating subject and worth understanding. Flowers produce nectar to attract bees as a reward to distribute pollen and aid in pollination. Nectar is produced as pollen matures and is ready for distribution by pollinators, and in some plants the volume of nectar produced increases with more pollinator visits. Floral nectar consists mainly of sugars and water in the ratio of 30% sugars to 70% water. The composition of the sugars varies with the floral species but usually a majority of sucrose, there are many other chemical components.

Forager bees land on a flower, or if it is too delicate, a stem and capture the nectar reward, unintentionally collecting pollen on the many hairs on its legs and body.

*The process of turning nectar into honey involves bacteria and enzymes in the honey stomach or crop.*

The process of turning nectar into honey involves bacteria and enzymes in the honey stomach or crop. Invertase, coverts sucrose to glucose and fructose and is added through the salivary and hypopharyngeal glands of foraging bees. On returning to the hive the foragers regurgitate the nectar and it is taken directly by a worker. Once passed from the forager to the house bees regurgitation takes place in a group, with bees partially digesting it through the addition of friendly bacteria and enzymes. This process also aids in evaporation, reducing the moisture content before storing the honey.

Lactic acid bacteria (LAB) an important group of gut flora are added that protect curing honey from fermentation. As the water content is further reduced the nectar is placed into cells where bees fan it with their wings to cause further evaporation. During the ripening process while the water content remains relatively high the enzyme glucose oxidase converts glucose to gluconic acid and hydrogen

*Prickly Box, Bursaria spinose, crystallises rapidly in the comb requiring extraction as soon as it is capped.*

peroxide that protects the ripening honey from fermentation by providing a high pH and antimicrobial activity. The ripening process continues for up to five days depending on humidity until the final product, honey reaches less than 18 percent moisture content, and it is then capped with wax for storage. The super-saturated sugar solution of honey provides natural resistance to infection, as it has moisture absorbing properties which dehydrates most microbes.

Ripe honey, the sticky sweet substance we are all familiar with, has a long shelf life and will not ferment if stored properly in sealed containers. "The oldest honey was unearthed in Georgia and was reportedly 5,000 years old (Lomsadze:2012)".

Crystallizing

Most honey crystallizes at some stage but the rate of crystallizing depends on the concentration of sugars. Honey with a high glucose to fructose ratio will granulate faster. Prickly Box Bursaria spinosa is a Tasmanian native species, common in eastern states, will granulate in the comb if not extracted as soon as 70-80% of a frame is capped. Its rapid crystallization indicates higher glucose content.

Canola is a crop grown in Tasmania, pollinated by bees, its honey also granulates rapidly. Crystallisation is a natural process caused by the formation of crystals of D-glucose monohydrate. It is more rapid in cold weather or when honey is stored between 10°C - 15°C.

Crystals vary in size due to the rate of crystallisation and produce fine to course grained crystallised or candied honey. However raw honey most often available by direct sales from beekeepers or at market stalls crystallises naturally, a good indication that it has not been overheated. Some commercial honey sold into the major retail trade is heated, cooled and filtered by packers to remain stable as a sweet liquid on supermarket shelves.

Historically liquid honey has been preferred by consumers and is still the majority of honey sold in supermarkets. Larger commercial producers and packers micro filter and flash heat the honey to 71°C then rapidly cool it. This has a number of affects, pasteurizing the honey, killing yeast cells and removing crystal forming nuclei such as fine wax particles and pollen. This provides the convenience of 'runny honey' and a long shelf life as the process of crystal forming increases the moisture content and the possibility of fermentation increases.

Overheating honey degrades it and produces increased levels of hydroxymethylfurfural HMF. This is a natural product of the decomposition of glucose and fructose in the presence of gluconic acid. Its production is increased with heat and with time. "Higher levels of HMF are an indication of damage from overheating, age of honey and may indicate contamination". The Codex standard allows for a level of less than 40mg per kg of HMF in honey from temperate and cool regions.

There is a growing awareness of the benefits of unprocessed honey, honey that has not been heated beyond hive temperature of 35°C. Eva Crane the renowned honey bee researcher notes that substances that give honey its aroma are volatile and heating honey above 30-35°C will degrade flavour and aroma (Crane:1990).

Hydrogen peroxide is the most common antibacterial element in honey produced by the bee enzyme glucose oxidase which is heat sensitive and reduces with time. Non-peroxide antimicrobial activity tested in Manuka, Leptospermum scoparium and commercially available as a medical grade product is not heat sensitive. This non-peroxide antimicrobial activity has been tested in some Tasmanian Manuka, Leatherwood and a number of Australian species particularly from the genus Leptospermum (Irish et al 2011).

*Honey is composed predominantly of the simple sugars, glucose and fructose, and water.*

Honey is composed predominantly of the simple sugars, glucose and fructose, and water, and it contains many other substances often in trace amounts, some coming from the bees' honey stomach and some from plant products.

General Chemical Composition of Australian Honey

| Composition | Range |
|---|---|
| Colour (Pfund Value in mm) | 32-78 |
| Moisture % | 15-18<br>Leatherwood 15.5 |
| Fructose % | 32-54<br>Leatherwood 43 |
| Glucose (Dextrose)% | 25-36<br>Leatherwood 30 |
| Sucrose% | 0.8-5.0<br>Leatherwood 2.4 |
| Maltose% | 1.7-11.8 |
| Nitrogen% | 0.5-.38 |
| Ash% | 0.04-.93 |
| pH | 3.3-5.6<br>Leatherwood 4.8 |
| Enzymes | Invertase, diastase, glucose oxidase |
| Acid | 0.5% (mainly Gluconic acid) |
| Free Acid | 12-40 m-equiv./kg<br>Leatherwood 14.7 |
| Vitamins | Minimal, less than 10% of Australian RDI |
| Minerals | Minimal, less than 10% of Australian RDI |

(Adapted from D'Arcy 2007) and (Arcot and Brand-Miller 2005). Leatherwood average from (Chandler et al 1974).

Australian Honey Colour

An official honey colour grading system has been adopted in Australia to align with international standards. It is based on the Pfund scale, a scale adopted by the honey industry that originally related to an amber coloured wedge of glass, the Pfund measured in mm this being the distance from the end of the wedge to the colour match.  Digital machines have now been produced that provide an accurate and consistent readout from honey samples.

*A great comparison of the various Tasmanian honeys. Tasmanian Beekeepers Association*

| Pfund | Official Australian Grade | Examples from Tasmania and Australia |
|-------|---------------------------|--------------------------------------|
| 0-34 | White | Blackberry, raspberry, clover. borage |
| 35-48 | Extra light amber | Brush box, lavender |
| 48-65 | Light amber | E. viminalis, (White gum)  E. pauciflora (Snow gum) Lemon,  Leatherwood |
| 65-83 | Pale amber | Blue Gum |
| 83-100 | Medium amber | Blueberry, Manuka |
| 101-114 | Amber | E. obliqua (Stringybark), E. amygdalina (Peppermint), Manuka |
| Above 114 | Dark Amber | Banksia, Paperbark, Buckwheat |

Thixotropy, Thixo what?

Manuka honey and some other Leptospermum species exhibit a phenomena known as thixotropy, a jelly like state that changes to a more normal liquid when vigorously stirred. When you lick a spoon of Manuka, it is hard to get it all off, and it also requires special extraction to get it all out of the comb. Thixotropy is caused by high protein content in the honey and can become more liquid on stirring or increasing the moisture content.

*The health benefits of honey have been used in traditional and folk medicine for thousands of years. Honey is one of the oldest medicines know, its recorded use going back more than 4 millennia (Molan:2007).*

Health Attributes of Honey

The health benefits of honey have been used in traditional and folk medicine for thousands of years. Honey is one of the oldest medicines know, its recorded use going back more than 4 millennia (Molan:2007). Its use is described in ancient texts, religious manuscripts and throughout historical literature. The first recorded use of honey as a topical wound treatment was from the writings of the early Egyptians in the Smith papyrus in 1650 BC (Mwipatayi et al 2004). The use of honey as a traditional medicine continues today being used to treat many ailments and wounds. Its use in modern professional medicine is increasing with more clinical trials being undertaken and medical grades of honey being produced. As more resistance is developed to antibiotics there is an increasing trend to using bio-active honey for wound care as it produces no adverse effects.

The ability of honey to provide healing comes from a number of known and many unknown qualities. The super saturated sugar concentration of high osmolality causes bacteria to dry out as honey absorbs moisture and also assists with natural debriding of wounds. The acid nature of honey with a pH between 3.3-5.6 also inhibits

*This ancient Egyptian manuscript provides the earliest evidence of honey being used in wound cure.*

infection. The natural occurrence of hydrogen peroxide in some honey is produced by the enzymatic action of glucose oxidase as the honey is being evaporated providing mostly sterile honey in the wax capped stores.

Glucose oxidase does not function at low moisture contents in stored honey but when honey applied as a dressing is diluted by wound exudates, the enzymatic production of hydrogen peroxide increases and is slowly released providing an antibiotic affect. The discovery of the Unknown Manuka UMF®, a non-peroxide antimicrobial effect was discovered by Professor Molan of the Waikato Honey Research Unit in New Zealand. Since that discovery methyglyoxal has been isolated as a non-peroxide antimicrobial when in honey. It is thought that the action of these substances relates to a combination of many phytochemicals within particular honey and most famously in Manuka, Leptospermum scoparium, and other Leptospermum sp. However to claim potency each batch of honey must be tested.

## Medicinal Honey in Tasmania

A group of the larger commercial honey producers in Tasmania have formed the Tasmanian Active Honey Group Pty Ltd. to further the development of the Tasmanian active honey sector. Tasmanian Manuka, a native species, is tested for its methylglyoxal content providing some indication of its bioactive potency. The group collaborates on further research into active Tasmanian honeys and has developed the Certified Tasmanian Active Honey Program, a quality assurance program to provide consumer confidence in Tasmanian active hive products.

## Other Health Benefits

Prebiotic : Research has identified that a number of Australian eucalypt honey's function as a prebiotic, stimulating the growth of gut bacteria that contribute to human health and reducing the growth of deleterious gut bacteria (Dawes and Dall: 2014). This may also apply to Tasmania eucalypt honey, but will require further research.

Work continues on identifying other health related attributes of honey such as its anti-oxidant value and low Glycaemic Index. Much of this work is inconclusive at the time of writing, but breakthroughs in any of these areas will tend to increase the value of honey.

## Caution

On a cautionary note do not feed honey to infants under one year old. Honey can contain the sporidium of Clostridium botulinum bacterium that produces toxins causing infant botulism in the undeveloped intestinal tract of children under one. This is extremely rare in Australia with only two cases (adults) of food borne botulism reported between 1999 and 2007. Clostridium botulinum is a widespread bacterium that inhabits rivers, soil, and the guts of mammals, fish, and shellfish.

In Australia there are no known cases of infant botulism attributed to honey. Botulism in infants and the unborn is a serious medical issue. Children over one year old and adults are not affected.

*Tasmania is a foodie's destination, it is on the global food map and our natural advantage is providing chefs and food entrepreneurs with a wonderful resource to produce outstanding food.*

Cooking with honey

Tasmania is a foodie's destination, it is on the global food map and our natural advantage is providing chefs and food entrepreneurs with a wonderful resource to produce outstanding food. Our honey is right at the forefront with "leatherwood honey being recognised for its wilderness harvest, its bouquet and flavour and increasingly its health benefits". With a number of unifloral honeys and wilderness wildflower honey available there are many opportunities for culinary development.

There is a wealth of recipes available for using honey and rather than fill these pages with recipes, a few keys to using honey are given and some indication of the use of our unique honeys. The rest is up to you.

Honey in cooking is effectively a sugar substitute, a natural sweetener that requires a slightly different approach. Honey is made up of simple sugars that are easy to digest. Honey in baked goods tends to darken them if

*Loukamades a Greek donut made with honey and walnuts. mygreekdish.com*

cooked at the same temperature as sugar, drop the temperature by about 15°C and cook it a bit longer, and you will need to experiment unless you have a specific honey recipe. Darker honeys tend to be more sensitive to overheating, most likely due to higher content of minerals and protein (Flottum:2011).

Using honey in baking naturally increases the temperature of the honey and causes loss of the volatile aromatics and heat sensitive enzymes. However, it is still a widely used source of a natural sweetener and imparts unique honey flavours while maintaining a moister body and a stickiness due to its hygroscopic property.

• Wipe your measuring spoon or cup with oil, this releases the honey

• Reduce the liquid in a recipe by ¼ cup for each cup of honey used

• Careful with the measure, honey is 1.5 times sweeter than sugar

• Acidity of honey can be reduced by adding ¼ teaspoon of baking soda per cup of honey used

(After Sammataro and Avitabile 2011)

*A batch of gingerbread dough uses a cup of honey*

Honey has been used as a culinary ingredient for thousands of years and continues to be used as both a commodity and a unique ingredient. Google honey or traditional honey recipes, they are almost without end. And now there is raw food, a genre where there is no cooking and the honey stays in its raw state. Explore, experiment and enjoy!

# CHAPTER FOURTEEN

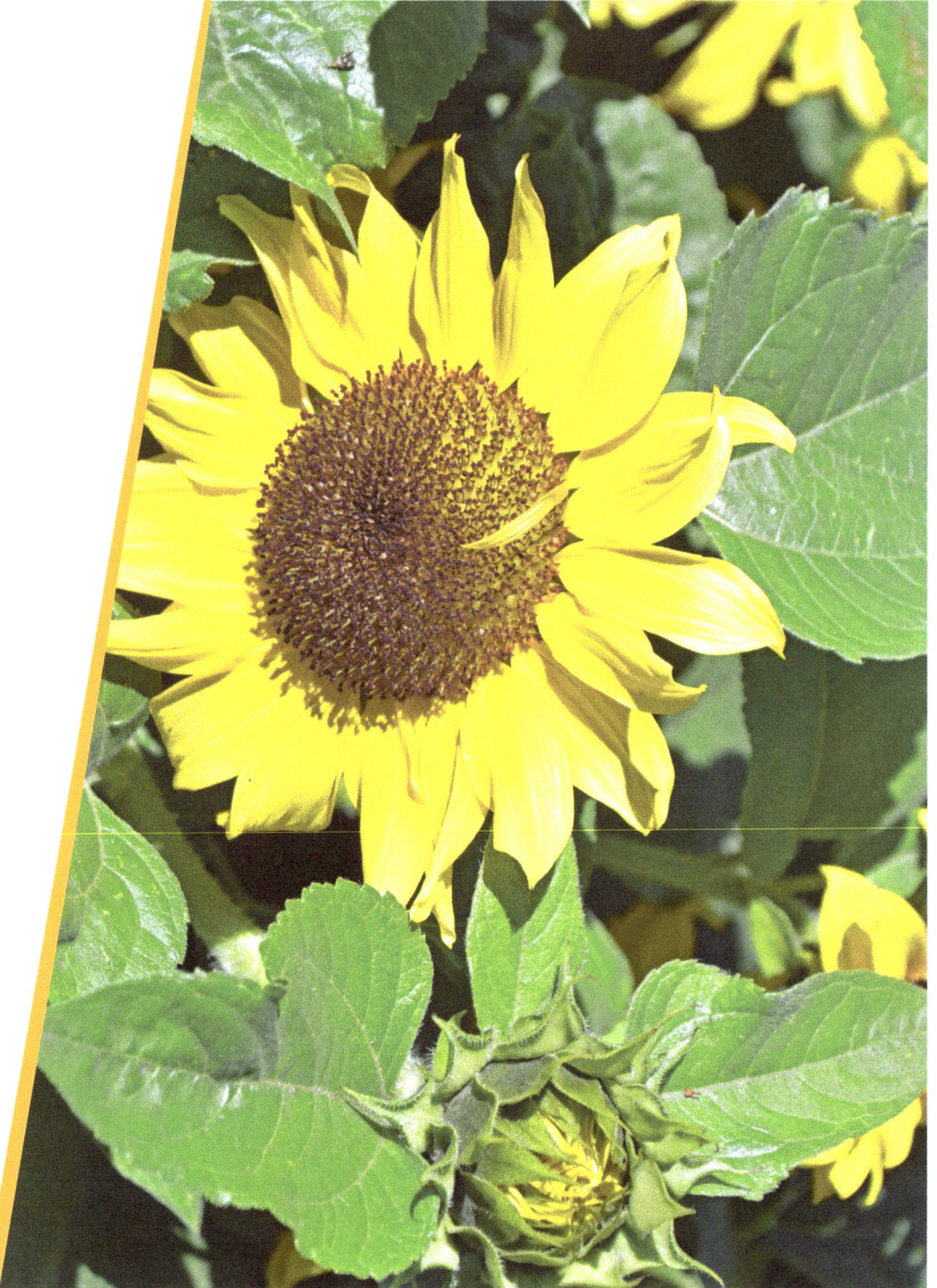

The selection of appropriate queens is fundamental to the success of your colony, and simple breeding procedures will provide an understanding of what to look for in your colonies behaviour in selecting the 'mother' queen. The key attributes that you should be considering are hygienic traits, non-aggressive behaviour and minimal swarming.

*Healthy queen David Barton Photography*

Commercial breeders and producers will be looking to optimise honey production, rapid build-up and other attributes at the expense of a happy hive but ideally we want it all. Commercial queen breeders may use instrument insemination, a more technical approach requiring a high level of skill and with known parents, queens and drones, breeders are then able to produce pedigree lines. The following will focus on some simple proven methods for raising queens.

*All queen breeding follows the simple principle that nurse bees can transform one day old worker larvae into a queen by enlarging the cell and feeding them for a longer period on royal jelly.*

All queen breeding follows the simple principle that nurse bees can transform one day old worker larvae into a queen by enlarging the cell and feeding them for a longer period on royal jelly. Queen rearing follows this principle by introducing a minute larvae, less than 24 hour old to a group of nurse bees in need of a queen (Cramp 2008).

Queens are raised by honey bee colonies in specially built queen cells. These are near vertical, peanut-shaped beeswax cells that have an opening that faces downwards and these queen cells may be found on the edges or surface of the comb. They are not always present, and they are only built as needed (DEPI Vic:2014).

Queens emerge from their specially constructed, larger, stronger, downward pointing cells 16 days after being laid and the wax of her cell is highly sculptured, stronger and she requires the help of worker bees to emerge.

Once emerged she will look for other queens or queen cells with viable queens that she will try and kill. Her sting is barbless and can repeat sting and is usually only used to kill competing queens. Often she will be the result of the hive having swarmed and workers preparing the hive for a new queen,

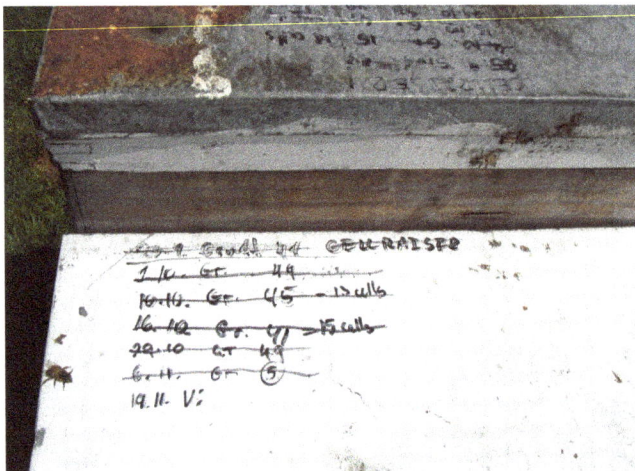

*Breeding notes indicating the source of the queen and time of introduction are often written on hive lids.*

in this case there is unlikely to be another queen present.

Newly emerged virgin queens feed themselves for 4 days until they start to produce queen pheromone making them very attractive to workers, who then feed them for the rest of their lives. At the same time they are sexually mature and ready to leave the hive to be mated by up to 20 drones over a number of flights. The sperm from this diverse group of male bees is kept alive in her spermatheca until needed to fertilise an egg. Once mated by enough drones she commences her lifelong task of egg laying.

Record keeping is essential to identify and remember desirable traits of your colonies. This aids in simple selection and is basic breeding practice. Information can be written on the hive lid or in a field note book or electronic tablet.

Importing Queens

It is a common practice to buy in queens from specialised breeders. Whether this is for a 'breeder queen' or all your queens, you must have the queens and her escorts inspected for small hive beetle.

On arrival in Tasmania the packaging and queen cages are to be inspected for evidence of small hive beetle by an Inspector. For bees coming by post this will be done at the Australia Post mail centre. If no small hive beetle is detected the package will be resealed and continue through the postal system. If small hive beetle is detected the whole consignment will be held and either destroyed under quarantine supervision or sealed and returned to the state of origin;

You the beekeeper must inspect the queens for small hive beetle prior to placing them in a hive. If there is any evidence of small hive beetle the queen cages must be sealed in a plastic bag and an Inspector notified. Bees or queen cages with evidence of small hive beetles must not be placed in a hive (DPIPWE:2014).

Colony Characteristics

The characteristics of the colony, its vigour, hygiene, temperament, tendency to swarm, honey production and many more qualities are determined by the genetics of the queen. Any or all of these can be changed by changing your queen. The most important characteristics for urban, backyard beekeeping are temperament and swarming tendency. Many commercial beekeepers replace their queens on an annual or 1-2 year basis as younger queens are more vigorous in their laying and emit higher levels of queen pheromone. Queen Mandibular Pheromone (QMP) a five chemical cocktail is produced in the queens mandibular glands. The chemical that says the queen is present, mated, young and active prevents the development of ovaries in worker bees and attracts her attendants. Decreased levels of queen pheromone is a signal to replace the queen, and build new queen cells, supersede or swarm.

Marking Queens

**International Code for Queen Marking**
Colour coding allows you to easily identify the year a queen was introduced to the hive. Mark the queen with a coloured paint marker on the back of the thorax. Queens do not usually live more than 5 years and it is a good beekeeping practice to regularly re-queen, thus maintaining a younger vigorous queen.

| Years Ending In | Colour |
|---|---|
| 1 or 6 | White |
| 2 or 7 | Yellow |
| 3 or 8 | Red |
| 4 or 9 | Green |
| 5 or 0 | Blue |

*Take care when handling your queen, hold her thorax between thumb and forefinger*

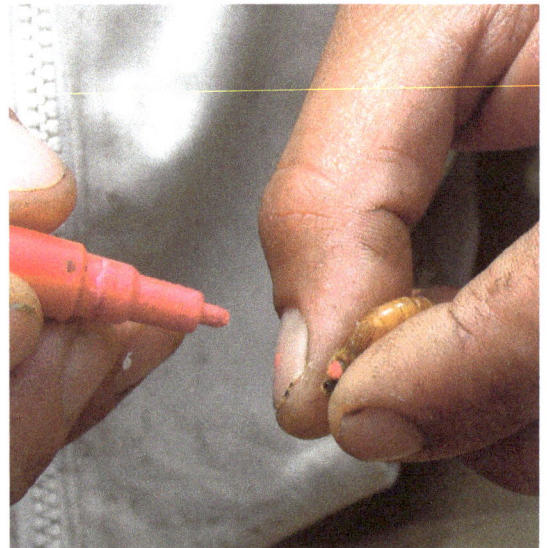

*An international colour code is used for marking queens*

Why mark your queens? If you are a beginning beekeeper this is an easy question to answer, even when marked they can be covered by their attendants and still hard to find.

Many breeders and commercial beekeepers mark their queens to make finding them easier as well as tell when the queen is absent and determine her age. An international colour convention has been developed for queen marking.

There are some disadvantages and risks to marking your queens. It is easy to damage the queen when handling her, at worst she could be killed or injury may affect her egg laying and cause supersedure. A number of devices or queen catchers/cages make it easier and safer for less experienced beekeepers to safely immobilise the queen while she is being marked.

Natural replacement, breeding or buying?

Bee colonies naturally raise new queens to replace a failing queen (supersedure), to replace a lost queen or to replace a queen from swarming.

Lost or dead queens are replaced by the reconstruction of several worker cells in the middle of the comb containing larvae less than 24 hours old. Worker bees tear the side walls down from several adjacent cells and construct a larger vertical queen cell around the young larvae. These larvae are then fed royal jelly all their larval life until the cell is capped.

*Supersedure cells are constructed near the middle of the comb*

*Queen cells are constructed at the bottom of the frame. Melbourne City Rooftop Honey*

Swarm cells are produced by a populous colony preparing to swarm, a survival response and the future queens are fed an abundant supply of royal jelly. The virgin queens usually emerge about two days after swarming and are bigger than queens produced by emergency replacement. Swarm cells are usually constructed around the edge of the comb or bottom bar and there may be as many as 10-25 queen cell cups. The wax in swarm cells is usually lighter in colour.

*Buying in queens from commercial queen breeders of proven well known lines is common practice, but all beekeepers need to have an understanding of queen breeding.*

### Buying or rearing

Buying in queens from commercial queen breeders of proven well known lines is common practice, but all beekeepers need to have an understanding of queen breeding. With a small apiary your ability to choose across colonies is limited, but you can still breed some useful queens. Whatever you choose to do, regular queen replacement every one to two years is essential to minimise swarming behaviour and optimise the vigour and health of your colony. Remember with queen pheromone decreasing by half each year, a two year old queen has a quarter of the pheromone of a new queen.

### Timing of requeening

The best time to requeen is during a moderate honey flow, usually in the swarming season, in Tasmania from November through December. However if it's a beekeeper problem where you have accidentally lost, injured or killed her it may be at any time and you will need to simulate a honey flow.

### Rearing

The following simple procedures can produce good queens and can be accomplished with practice. To successfully rear queens you need the right conditions; warm weather, a light nectar flow, at least three sources of pollen, a hive containing a large number of bees, brood being produced, a large number of drones and queen cells being constructed.

### Raising queens from hive produced queen cells:

Select a hive with desirable characteristics (a bit hard if you only have one or two hives), and a method to produce five or six queens from queen cells that are full-sized, sealed and have 'pared' tips. These queen cells may be carefully removed just before or in the course of swarming and transferred to an equal number of nucleus hives. From the nucleus hives queens emerge and mate within six or seven days of the date of transfer.

Breeding is not controlled, your queen will have mated with up to twenty unknown drones, but it is hoped that some of her traits will be inherited in your new queens.

A nucleus hive can be created by removing three frames of bees, sealed brood and stores from any good hive, and placing them in a special three-frame box or hive body with a contracted entrance. The 'ripe' queen cell placed towards the top of the middle frame. Preferably the cell should be in a wire cell protector.

Key points to follow:

- Ensure there is no queen or queen cell in the three frames selected.

- The 'ripe' queen cell should be lightly held by its broad wax base and kept in its inverted position while it is transferred, insert in the cell protector and fix to the centre frame of the nucleus hive.

- Do Not Disturb! Apart from re-opening the entrance on the third day, nucleus hives should not be disturbed or examined for eight days after the introduction of the queen cell. At the end of this period the hive may be examined quickly and carefully using a minimum of smoke.

*A cell bar full of queen cups and constructed cells.*

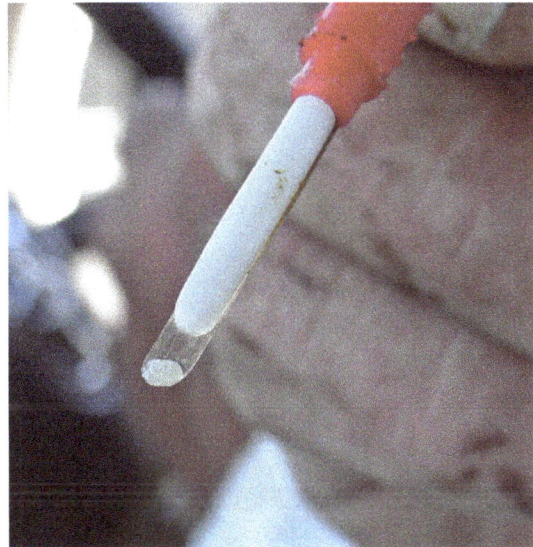

*A fine grafting tool*

Grafting:

Grafting very young larvae into artificial queen cells may be a little advanced but it is not beyond enthusiastic amateur beekeepers and is worth practicing especially if a few of your friends want to share the queens you produce. You may even be able to sell your nucleus hives or create new hives for other beginners.

You will need some simple equipment: artificial wax or plastic queen cups, a cell bar that the cells are attached to with hot wax, a grafting tool, a number of three to five frame nucleus boxes for starting cells and mating queens.

*Grafting being undertaken by a master beekeeper*

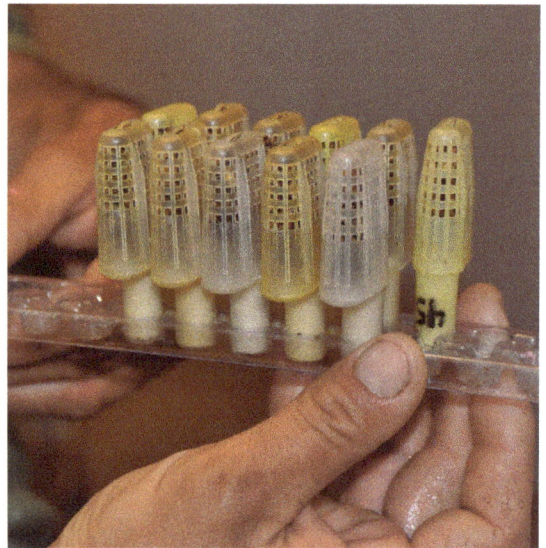

*Plastic queen cages*

Prepare a queenless nucleus to accept the grafted larvae as they will go into emergency queen replacement mode.

Introduction of caged queens

Before introducing your new queen, either one you've purchased or bred yourself, you need to find and remove the old queen. You will also need to remove the attendant bees in the cage. The best time to re-queen is during a honeyflow.

It is important to observe the reaction of the hive bees to the introduced cage. If they are clustering around her and trying to lick and feed her, all is well. However, if they are trying to bite her it may indicate that another queen is already present

*A queen cage preloaded with candy and gently loading the queen and some attendants*

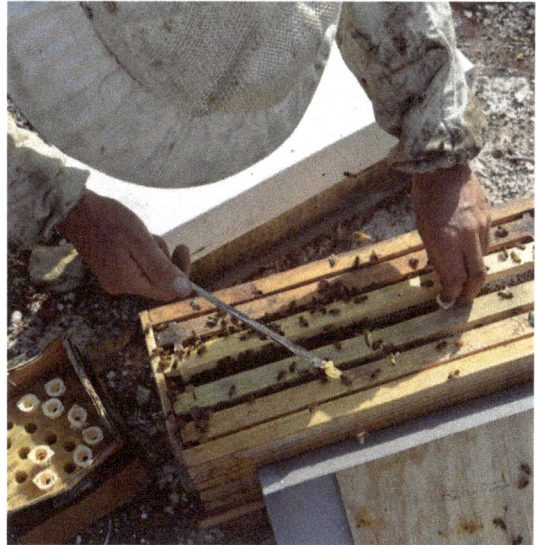

*Loading queen cells into a nucleus hive, the start of a new colony*

in the hive. Check back after a week and look for the presence of new brood indicating her acceptance and that she is busy laying.

You can now create splits, increase the size of your apiary and breed queens, well done!

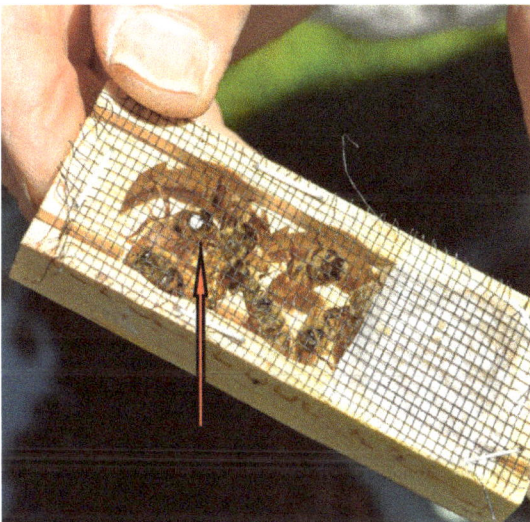

*A wire "transport-mailer" queen cage. A young queen with attendant workers*

*Inserting a mailer cage into a nuc*

*Four nucs from an innovative commercial beekeeper*

*Nuc hives with mated queens M Leech*

**A Queen Rearing Program**

The following queen rearing program is taught at the annual southern beekeeping course. It is important even for beginner beekeeper to gain an understanding of queen rearing and to become confident in raising your own queens.

| Day | Task to be undertaken |
|---|---|
| Day 1 | Prepare cell raising hive<br>Pick a strong disease free hive, this should consist of at least 5 ideal or 2full-depth supers full of bees with abundant brood. Find the queen and ensure that she is in the bottom half of the hives and place a queen excluder above her. |
| Day 3 | Frame in breeder<br>Place an empty frame of dark wax (easier to see larvae and wax is harder, Cowan 2015) or light wax in the hive you are going to breed from. Restrict the queen to one full depth or ideal super. This will give you the right aged larvae to graft. |
| Day 7 | Condense cell raising hive and insert cell cups to be cleaned.<br>Take top ideal super off to make bees tighter, not applicable to full depths.<br>Attach queen cells to a cell raising frame and place into the centre of the top super for the bees to clean and polish.<br>Give the hive a small amount of slightly warm syrup if no nectar coming in. |

| | |
|---|---|
| Day 8 | **Grafting** <br> Take cells on bar out of the cell raiser (leave the gap) and graft larvae from the frame placed earlier in the breeder. Place the cell bar back in the gap. Feed again. Bees will cluster in the gap and attend to new larvae immediately they are introduced. |
| Day 9 | 24 hrs after grafting you can tell if they have been accepted. They will have begun to build out cells and the larvae will be floating in royal jelly. |
| Day 16 | Establish enough nuclei hives for queen cells <br> A 4 frame nucleus should consist of 2 frames with honey and 2 frames of brood, some unsealed, hatching and capped and the brood should be well covered by bees. If a full super then double these figures. Feed sugar syrup if no honey available. <br> Its very important for good queens to be well looked after when they hatch. Leave the nuc locked, but ventilated in a shady place for a minimum of 32hrs prior to moving to the mating yard. Make sure that the nuc you picked is queenless. |
| Day 17 | Take the nucs to your mating yard and let out in the evening. Keep nucs at least 2m apart facing different directions, even use different colours to make it easier for the queen to locate her new home. |
| Day 18 | Take cells from the breeder hive and place one in each nuc. <br> Pry off the cell bar with a knife and place in a queen cell protector. Pick the biggest most sculptured cells. These will be the strongest queens **Do not tip cells upside down while handling** as queen wing development is very fast and still developing while placing the cell into the nuc. |
| Day 23 | Check whether cells have hatched. <br> **These nucs cannot be moved until the newly mated queen is laying. This can take 10 to 20 days.** <br><br> Wait until she has been laying for three weeks before introducing her |
| | **Records** <br> The date of grafting can the written on the cell bar with a breeder code. The nuc can have the hatching date and breeder code and the final hive can have the breeder code and the date introduced. This latter can apply to purchased queens using a simple code for the breeder. |

The honeybee diet consists predominantly of three food groups, carbohydrates, protein and water. The aim of beekeepers chasing honey flows or urban beekeepers wanting to optimise honey yield and for crop pollinators is to have strong colonies with a maximum number of foragers or field bees. To achieve this you need healthy colonies with good bee forage or supplementary feeding at the right time.

Bees require a varied diet to maintain their health, vigour and resist disease. Research that compared feeding a single nutritious pollen to a multi-pollen diet showed that polyfloral diets can provide enhanced immune function in honeybees with an increase in the enzyme glucose oxidase, responsible for the production of hydrogen peroxide. The availability of floral diversity could provide enhanced in-hive anti-microbial activity and disease resistance (Alaux et al 2010, Huang 2010).

*Early flowering plums provide much needed nectar and pollen*

*Vast areas in the US and increasingly Australia are almond monocultures. Almonds do have high quality protein, but there's no variety in the diet for weeks*

The notion of floral abundance and healthy bees is challenged with a reduction in biodiversity across the landscape. Larger areas of monoculture planting and the removal of remnant vegetation to improve irrigation and cultivation efficiency is a major concern.

It is considered by many that a major contributing factor to the widely reported

*Salvia leucantha is a very beneficial bee plant, long flowering summer to winter. In its natural environment in the US it produces the famous Sage honey*

syndrome, Colony Collapse Disorder is a reduction in the quality and diversity of bee diet from a decrease in floral diversity (Mussen 2009) and (Heintz 2009).

Many authors say the most important design criterion for a bee-friendly domestic garden is floral abundance and continuous flowering throughout the year (Somerville 2002; Shepherd 2004; Barrette 2010; Goodman 2010; IBRA 2008). Bees in urban colonies have the benefit of gardens, with year-round flowering from many different species, different foraging habitats, and a variety of pollens and nectar (Leech 2012). This often ensures a well balanced diet not lacking in any essential elements.

Several publications provide more detailed information on honeybee nutrition, reviewing the science and identifying gaps in the knowledge base including (Black

*Bees in urban colonies have the benefit of gardens, with year-round flowering from many different species, different foraging habitats, and a variety of pollens and nectar (Leech 2012).*

2006). The RIRDC publication 'Fat Bees Skinny Bees', (Sommerville 2005), is recommended as a manual on nutrition and how, what and when to artificially feed your bees.

Bee Food

Bees mainly get their carbohydrate requirements from nectar produced by floral nectaries and the protein from pollen.

Nectar:

- Provides carbohydrate and is the main energy source for bees.

- Mainly consists of water and sucrose and smaller concentrations of glucose and fructose. (do not confuse with honey)

- Composition varies greatly between plant species.

Pollen:

- Pollen is the protein source for bees and is one of the main nutrients for growth.

- Pollen contains, amino acids fats, fatty acids, sterols, carbohydrates (fibrous material) vitamins and minerals.

Pollen content and composition varies widely with plant species and is not constant but is affected by growing conditions; soil moisture, fertility and ambient temperature (Herbert 1997).

Pollen and Growth

Pollen consumption relates closely to the growth of the bees and development of the hypopharyngeal gland and peaks at 8-10 days after emergence and declines after 15 days once the worker commences foraging duties (Crailsheim and Stolbey 1989).

The crude protein content of pollen is calculated by measuring the nitrogen and multiplying it by a factor of 6.25. This measurement of nitrogen is inconsistent between laboratories and can provide a false lead when looking for forage.

*An environmental weed that is a bee magnet providing highly sought after pollen in late autumn/spring*

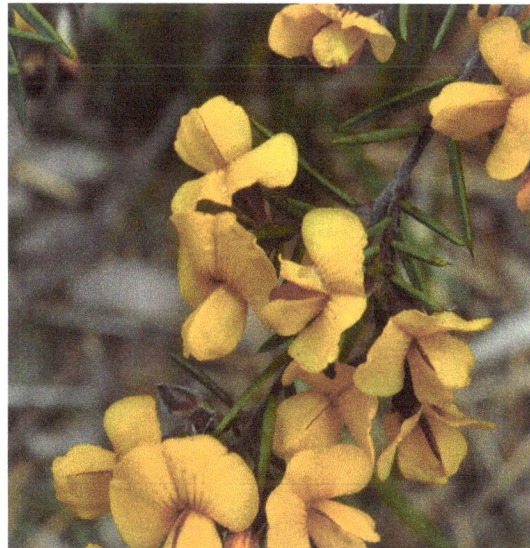

*These spring flowering understroy plants provide highly nutritious pollen*

# CHAPTER FIFTEEN

Proteins are composed of a series of amino acids. In 1953 De Groot discovered that honeybees have the same 10 essential amino acids as mammals and that there are minimum levels essential for honeybee development.

| | Minimum Levels of Essential Amino Acids for Honeybee Development | | | | | | | | | |
|---|---|---|---|---|---|---|---|---|---|---|
| Amino Acid | Arginine | Histidine | Lysine | Tryptophan | Phenylalanine | Methionine | Threomine | Leucine | Iso-leucine | Valine |
| g/16g trogen | 3.00 | 1.5 | 3.0 | 1.00 | 2.5 | 1.5 | 3.0 | 4.5 | 4.5 | 4.0 |

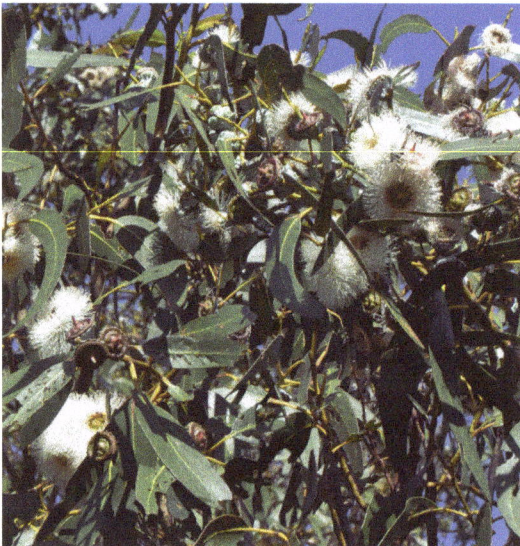

*Large blue gums provide an abundance of flowers about every 4 years with some flowers most years*

The composition of pollen is not constant for any plant species, but varies from site to site and year to year with many factors affecting growing conditions including soil moisture, fertility and ambient temperature (Black 2006).

Local knowledge suggests that some plants produce both highly nutritious and or abundant pollen. In Tasmania colonies are known to build very well on a number of environmental weeds including crack willow, Salix fragilis, cape weed, Arctotheca calendula and gorse, Ulex europea, while bees working on a Prickly Box Bursaria spinosa honey flow will be in prime condition to take to a leatherwood, Eucryphia lucida honey flow. This anecdotal information is in part supported by science, but two species, crack willow and cape weed are lower in crude protein

*This is a very bad environmental weed. It does provide abundant pollen that is high in crude protein*

*Prickly box is a primary coloniser of poorer sites like this rocky ridge*

than recommended levels. A factor overriding crude protein availability and lack in any of the 10 essential amino acids is pollen abundance and availability and importantly digestibility of the pollen. A lack in one of the essential amino acids would mean the colony has to make up for that lack in volume of pollen consumed. If the deficient amino acid is only at half the minimum level then the twice as much pollen would be needed.

Management of nutrition

The best management is to have your bees in an area with abundant flowering of multiple species from early spring until autumn. This may not always be possible and the following briefly describes the management of colony nutrition in the absence of floral abundance.

Your colonies can be rescued, enhanced and manipulated by changing their nutrition.

Stimulating colony growth is very important in Tasmania due to our short season. Growth

*Pultanaea Juniperina one of the egg and bacon species and a useful source of pollen*

| Some Valuable Tasmanian Species | | |
|---|---|---|
| **Species** | **CP%** | **Amino Acid Issues** |
| Gorse<br>Ulex europaea | 28 | Non limiting no Tryptophan |
| Crack Willow<br>Salix fragilis | 14.8 | Non limiting |
| Cape Weed<br>Arctotheca calendula | 16,75-21 | No Tryptophan |
| Wild Raddish<br>Raphanis raphanistram | 25.2 | Iso-leucine may be limiting |
| Lupins | 28-34.7 | Lysine maybe limiting |
| Boobyalla | 24.9% | Non-limiting |
| Pultanaea | 31-33% | |
| Blue Gum<br>Eucalyptus globulus | 27.6-29.6 | |
| White top<br>E. delegatensis | 23 | Maybe non limiting |
| Brown Top Stringy<br>E. obliqua | 24.3 | Non-limiting |
| White gum<br>E. viminalis | 21.3-23.7 | Arginine maybe limiting |
| Pride of Madiera<br>Echium candicans | >30 | |
| Saw leaf banksia<br>Banksia serrata | >31 | Non-limiting |

can be stimulated by feeding sugar syrup in spring or at other times. This is the fastest way to stimulate brood production and increase the need for pollen. This method is used by crop pollinators to artificially stimulate their colonies to forage for pollen and nectar when they would not normally. It rapidly builds the colony, providing more foragers to undertake the job of crop pollination while gathering food. Stimulating with sugar syrup has the same effect as a nectar flow and invokes hygienic behaviour in the colony.

The most common cause of colony death in winter is starvation. Feeding granulated sugar is a way of keeping the colony alive without stimulating brood production. The brood is kept at 35-36 °C and to maintain this through winter requires considerable energy using up valuable stores. A rule of thumb in Tasmania is to leave on two ideal supers or one full depth super of stores to keep the hive through winter and into spring flowering.

*A rule of thumb in Tasmania is to leave on two ideal supers or one full depth super of stores to keep the hive through winter and into spring flowering.*

Winter hives without brood cluster are their most efficient at 7 °C, clustering and using minimal stores (Somerville:2005). Hive stores should be checked at the beginning of winter and mid-winter by lifting up the hive - to estimate weight and get a feeling for remaining stores. You should avoid opening the hive in winter and definitely not open it on cold or wet days as it disrupts the cluster and the bees use considerable energy to bring the temperature of the brood cluster back to 35-36 °C. If you have to feed your bees in a dearth use an inside feeder to prevent robbing.

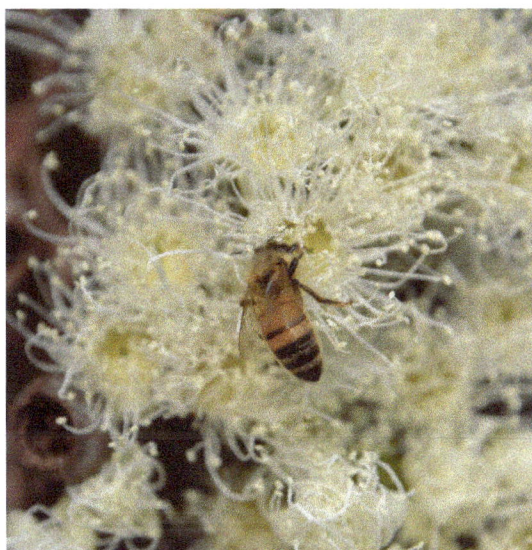

*Smooth barked apple, Angophora costata, is an excellent source of abundant good quality pollen and some nectar.*

*Lavender is an excellent bee food*

Pollen availability will affect the size of your brood and then your colony. Large healthy colonies are the aim of beekeeping with a focus on honey production. A lack of quality pollen will create a reduction in brood area. Pollen that's low in quality crude protein may be overcome with volume, but if one of the essential amino acids is below required minimum levels this will cause a reduction in brood. If however a pollen high in crude protein is deficient in one or more amino acids the deficiency may be overcome by increased pollen collection.

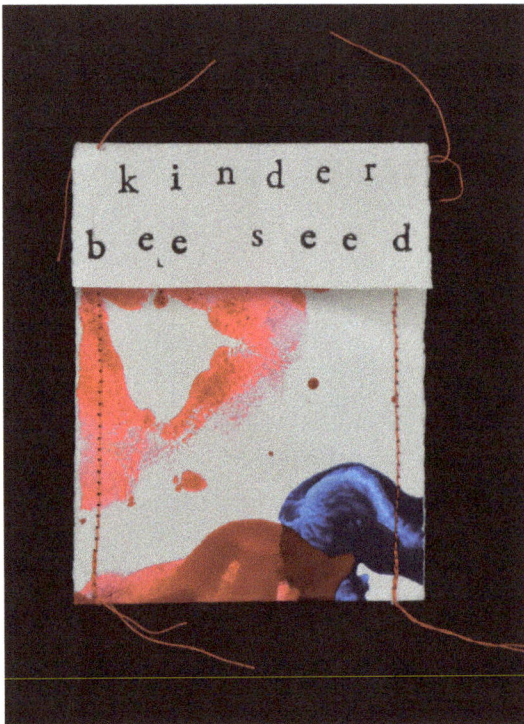

*A school project developed by a passionate kindergarten teacher. The children produced the artwork packs and collected and dried seed from bee friendly plants. This will make a difference!!!*

Pollen supplements are available from beekeeping suppliers and beekeepers can make their own. It is important to ensure pollen supplements are GMO free. Commercial beekeepers exporting to the EU cannot risk GMO contamination as it will affect sales. Supplement infers there is some pollen available as it is very rare that no pollen exists. Supplements and DIY recipes usually contain 10-20% pollen, 20-100% soy flour, 20-25% yeast, 20-50% of honey, sugar, water. The hardness of the patty supplement produced will be determined by the amount of sugar used, as more sugar provides a harder product. Bees seem to prefer a moister product, perhaps the consistency of marzipan. Flat pollen supplement patties can be placed across the top bars of brood frames.

Small apiaries of 1 to 20 hives located in a town or close to an urban area are unlikely to suffer a major pollen shortfall except in exceptional circumstances, given the wide variety of plants flowering at any time (Somerville 2005).

Consider planting for bees and encourage others as well. Plant near apiaries for a late winter early spring pollen and nectar source with a pollen emphasis so they can take advantage of short breaks in poor weather providing pollen for a growing colony. Close nectar sources will be helpful for autumn providing winter stores.

We can all make a difference, the time to plant is now!

Managed European honeybees are the most important insect in the pollination of the world's crops, and pollination by these bees is an accidental outcome of their search for food.

It is very important for beekeepers to understand that both feral bees and native bees assist honey bees in pollinating crops. While feral bee populations are the most at risk from an invasion of Varroa mite, enhancing native bee populations is a very important factor for the future. A key issue will be to better understand their habitat needs, their ecology and proactive engagement in this process. While this is not the main job of beekeepers, it is important to improve our understanding of the role and requirements of native bees.

*Commercial cherries are highly dependant on pollination from managed honeybees*

*Bees have to contend with being under nets for a number of crops*

The provision of pollination services by beekeepers to horticulture and agriculture is a significant business globally, and increasingly important in Tasmania as more high value horticultural crops are established and the production of seed crops for export is expanded.

Managed and wild or feral populations of European honeybees provide pollination services to an estimated 65% of horticultural and agricultural production systems in Australia (Keogh, et al., 2010). "Reliance on feral, unmanaged honeybees for pollination is a high risk management strategy" and better and more reliable pollination can usually be achieved by introducing managed colonies. Unmanaged feral colonies will be wiped out by a Varroa incursion. Beekeepers and growers need to establish strong relationships and better ways to engage.

The Tasmanian Crop Pollinators Association have pioneered the use of pollination contracts to establish the responsibilities of beekeepers and growers in the provision of pollination services. Communication between parties needs to commence early and be maintained. Beekeepers need to begin preparation of their hives well in advance of pollination jobs to ensure bee populations are optimal for foraging for pollen providing an effective pollination service. Growers need to be aware of bee health and modify or halt spraying regimes while bee colonies are present.

Tasmanian and Australian horticulture increasingly relies on managed honeybee pollination services and around the world and both wild and managed honeybee populations are under increasing pressure from changing environments, agricultural practices, pests and diseases.

*Managed honeybees pollinate canola to increase oil seed yields*

While commercial pollination provision is generally the work of larger commercial beekeepers, there are niche opportunities for smaller sideline operators and backyard beehives can provide exceptional outcomes in gardens and the urban environment.

Tasmanian crops that are highly dependent on bee pollination include high value seed crops: carrots, canola, brassicas, onions, lucerne and clover; fruit and berry crops such as cherries, apples, apricots and raspberries. The demand for pollination services is increasing with increased planting of pollination dependent high value horticulture and seed crops.

Management of pollination services often requires specific hive management strategies that can reduce the opportunity for maximum honey yield. The Tasmanian Crop Pollinators Association has developed a Code of Practice to assist in the orderly and professional management of crop pollination.

Pollination Essentials

Pollination means the transfer of pollen from the male part of the flower, the anthers, to the receptive female part, the stigma. Fertilisation occurs when the

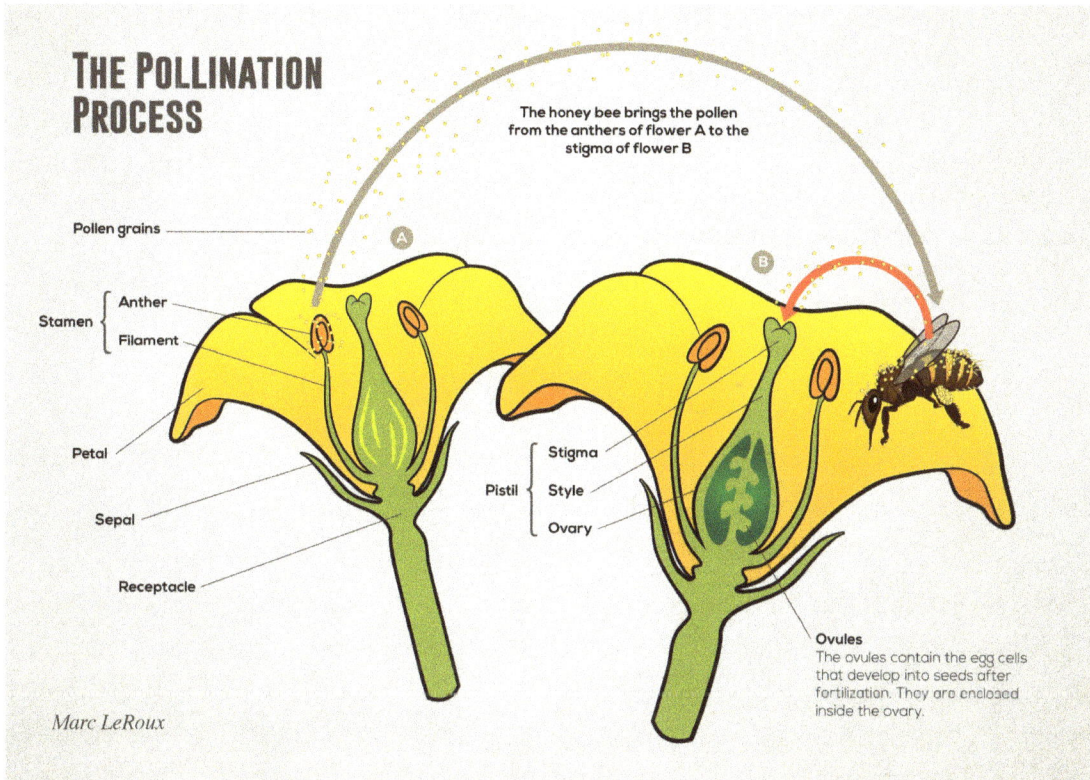

## THE POLLINATION PROCESS

The honey bee brings the pollen from the anthers of flower A to the stigma of flower B

Pollen grains

Stamen { Anther / Filament

Petal

Sepal

Receptacle

Stigma
Pistil { Style
Ovary

Ovules
The ovules contain the egg cells that develop into seeds after fertilization. They are enclosed inside the ovary.

*Marc LeRoux*

pollen grains on the stigma germinate and grow down the stem of the stigma (the style). The sperms of the pollen unite with the ovules in the ovary of the flower and subsequently produce seed. Pollination is a pre requisite to the fertilisation of ovules within flowers that leads to the growth of seeds and fruit, and the higher the level of effective pollination, the greater the size and yield of the crop (Tasmanian Crop Pollination Association: 2012).

Seasonal development of a hive

As discussed in chapter ten Seasonal Management bee populations build as food becomes available and weather warms. This is a simple equation that works fine for the amateur beekeeper and even commercial beekeepers managing for honey flows, but it does not provide adequate bee populations for pollinating early crops.

For all the early spring flowering crops such as almonds, cherries, apples, apricots, plums, and currants the bee populations have to be artificially fed and skilfully managed to reach optimum size and proportion of workers to brood.

The TCPA Code of Practice notes the following essentials for pre-pollination hive preparedness.

- Queen: A healthy, vigorous laying queen is necessary to ensure optimum egg laying, presence of young larvae and an adequate number of foragers.

- The minimum strength of a colony suitable for pollination at the end of winter or early spring is one that has brood on both sides of six ideal frames, with 50% open brood in all stages of development, (four frames Full Depth) with bees covering at least ten frames.

- Adequate stores of honey and pollen.

- Adequate honey stores prevent starvation when nectar is scarce and or with weather conditions that are unfavourable for foraging.

- Stores of pollen will encourage brood rearing.

The Code of Practice is essential reading for any beekeeper and grower contemplating commercial pollination. It provides details on issues of delivery times, quality of hives, safety for workers, spraying by growers, and includes annual pricing. It establishes a standard and performance requirements and includes an agreement to be signed by both parties.

*All the seeds in an apple need to be fertilised to produce fruit of good shape and size*

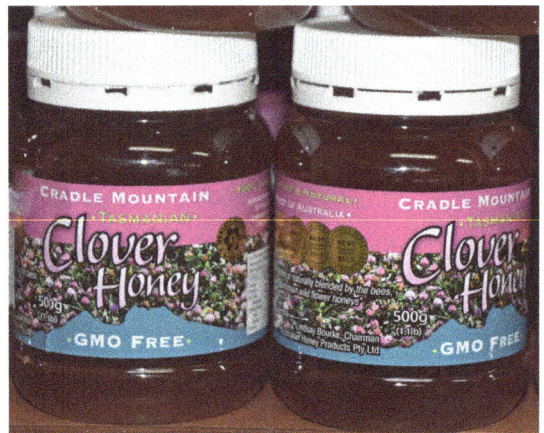

*It is essential that Tasmanian honey remains GMO free*

Tasmania's GMO Free Status

Tasmania's Genetically Modified Organism (GMO) free status is critical for the apiary industry and continued servicing of prime export markets. Honey cannot be exported to the European Union if it is within five kilometres of a GM crop. Tasmania's brand of pure, clean, wild and high quality must be maintained.

The beekeeper has many jobs to undertake and perhaps the most important is the monitoring of the health of the colony, and identification of any disease or pathogens that are present. This is not just a seasonal issue but a year round constant. Disease prevention is always better than intervention and can generally be achieved by ensuring you only use disease free equipment, regularly replace your queen ensuring she carries hygienic traits, replace discoloured brood comb on a systematic basis, keep plenty of stores, pollen and honey on over winter.

Location of the hive is also critical in helping to prevent disease; site your hive so that it is above dry ground, is protected from prevailing weather, not in continual shade and is near a clean water source. Over-robbing depletes the bees' natural storage of winter and early spring food used to maintain the colony when there is a lack of winter, early spring flowering and weather conditions that are adverse. Bees do not fly in wet cold weather and remain in the hive consuming stores.

*Avoid opening your hive through the winter and especially not on cold wet days as this may chill and kill brood.*

*An easily accessible, sunny well drained apiary site*

*A frame of healthy brood Karla Williams*

Avoid opening your hive through the winter and especially not on cold wet days as this may chill and kill brood.

It must be remembered that bees are always dying of natural causes and given that they only live for about 6 weeks in the active season, many bees die.

Colony Collapse Disorder CCD

Colony Collapse Disorder (CCD) a new name that emerged with unusually large colony losses (30-70% of hives) in the US in the winter of 2007 is now known throughout the western world and beyond, thanks to popular media. This reported plight of bee populations has ignited people's interest in the health and wellbeing of honeybees. Note this syndrome is not present in Tasmania or the Australian mainland.

There have been vast amounts of research to identify the causes of CCD, the general consensus now being that rather than a single disease or poison it is a syndrome caused by many different vectors in varying combinations. Increasing industrial agriculture and a build-up of possibly benign chemicals that in combination reduce honey bee's resilience to disease. However, there are a group of chemicals under moratorium in the EU, the neonicotinoids, a group of systemic chemicals used to coat seeds to make pollen toxic to unwanted insects, but significantly affecting beneficial ones like honeybees.

Modern large scale agriculture tends to decrease the diversity and abundance of forage. The increased reliance on honeybees as pollinators creates added pressure on bee populations especially in large areas of monoculture such as almonds.

While almond pollination may be beneficial as the only high quality pollen source for many weeks it could cause nutritional imbalance, notwithstanding the huge biorisk of bringing large numbers of hives across great distances in the US and Australia. This will never be at the same scale in Tasmania, but we must remain aware and look to maintaining bee forage diversity across the landscape.

Causes of loss

Bees are continually emerging as new adults and dying of natural causes. Worker bees tend to live for 6 weeks in the summer season as they wear out from flying, get damaged wings, or die of exhaustion. Bees that are fortunate enough to be born into a winter hive tend to live for 3-4 months and are the first foragers in the early warm days of spring. A colony with a vigorous young queen, abundant good forage and a well located hive, sheltered from prevailing winds, and not in a damp or shaded site will generally be healthy.

If you have a new hive, new equipment and new bees, which is well located and with good forage it is most unlikely that disease or disease related issues would show up in the first year. This may be different if you have acquired second hand

**Possible causes of loss are in the table below and explanation of the issues follows.**

**Diseases of adult bees -**

- Nosema apis
- Nosema ceranae
- Other

**Diseases of larvae –**

- American foul brood   AFB
  (Paenibacillus larvae subsp larvae)
- European foul brood EFB
  Mellissococcus pluton)
- Chalk Brood
- Sacbrood

**Human agencies –**

- Chemical sprays
- Vandals
- Poor beekeepers

**Pests and predators –**

- Braula fly (Braula coeca)
- Small Hive Beetle
- Varroa destructor
- Acarapis dorsalis
- Wax moth
- Birds, ants, mice

**Natural causes-**

- Starvation
- Pollen infected with mould
- Showery weather in spring
- Cold and exposure
- Chilled brood
- Overheated brood

Modified from Ayton 1989

equipment, not recommended for amateurs, as it can often come with disease or at least disease spores that can manifest in the right conditions. It is important that you are familiar with the types of disease and issues that may adversely affect your colonies. This may be as simple as not enough food and the symptoms of starvation or a more complex disease scenario.

Diseases of adult bees

Nosemosis, Nosema:

Caused by: Nosema apis and Nosema ceranae, Microsporidia.

What to look for? It is difficult to detect.

Nosema is the most widespread adult bee disease in the world. (Sommerville 2007)

Nosemosis (Nosema) is a disease of adult bees caused by microsporidia, a fungal spore forming single celled parasite. There are two known bee infecting Nosema species,  Nosema apis and Nosema ceranae. N. apis found in all states in Australia and N. ceranae is now in all states except Western Australia.

Nosemosis attacks the midgut of the bee and causes them to stop eating by reducing the digestion of pollen. It is also known to reduce the function of the food glands (hypopharyngeal) resulting in poor brood rearing ability as nurse bees cannot produce enough brood food.

Greater numbers of worker bees become infected than drones or queens, most likely due to comb cleaning activities as the disease spreads through faecal matter. Also, Nosema infected bees do not attend the queen to the same extent as healthy bees buffering her from infection. If a queen becomes infected her ovaries degenerate reducing her egg laying capacity that can cause supersedure. (Wikipedia Modified).

Infection is most likely when bees are stressed and in spring weather that is cool and damp, when re-queening and moving hives.

*Diarrhoea at the entrance can be a sign of Nosemosis*

## Symptoms:

Symptoms are confusing and there are no specific symptoms related to this disease. (Sammataro and Avitabile 2011, Goodman 2014) They can come from other causes such as dysentery and chemical poisoning in the case of confused behaviour.

Diarrhoea and sick bees crawling about the hive and on the grass in front of the hive and can be dragging their legs as if paralysed are correlated to Nosemosis. The only certain diagnosis is to examine the bee mid-gut. A healthy gut is tan coloured and wrinkly, and a Nosema infected gut is swollen, smooth and white. Symptoms include:

- Distended abdomens.

- Not eating when fed sugar syrup.

- Beekeepers becoming aware that their hives have a lot of brood and few bees.

- When a hive recovers from this setback it usually remains relatively free from the disease; but in severe cases the hive dies as sick bees staying in the hive reinfect through their spore laden faeces.

## Treatment

Production hives cannot be treated with the antibiotic fumagillin. Fumidil B can only be used in Tasmania and under permit in queen rearing operations. It remains as a residual chemical in honey.

Prevention

Ensure your bees have a continuing healthy diet with plenty of available stores. Pollen, honey and good forage for both nectar and fresh pollen, do not rob too much in autumn.

Hornitzky 2007 demonstrated that pollen availability is a key factor in bee longevity. Bees with a good pollen diet live longer, even if they are heavily infected with N. ceranae or N. apis. To prevent this;

• Re-queen regularly.

• Stimulate eating with essential oils if infected, a non-conventional but effective method.

• Replace old discoloured brood comb on a regular basis.

• Place hives in warmer, sheltered, not shaded or damp sites.

• Ensure hives sit off the ground on some form of hive stand such as purpose built stands, cement blocks and pallets.

Other

There are a number of symptoms; an apparent viral disease causing significant death with no known treatment that may be associated with Nosema, starvation, exposure, cold damp conditions and the possibility of poisoning. If you find large numbers of dead bees without a readily identifiable cause having consulted other beekeepers, contact the Government Apiary Specialist.

Diseases of Brood

American Foul Brood (AFB),

Caused by: Microbial disease by spore forming bacterium, Paenibacillus larvae subsp larvae. It is a destructive brood disease, does not affect adults.

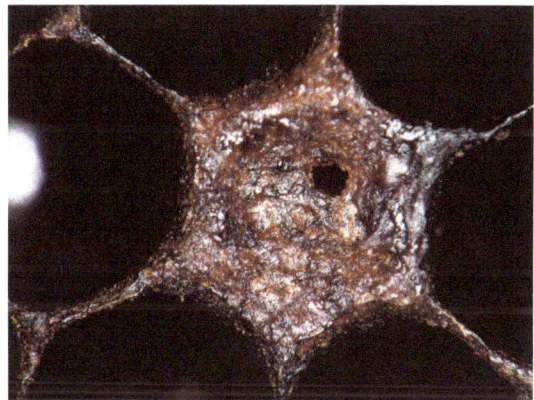

What to look for: In brood frames, larvae that are not pearly white, irregular brood pattern, sunken greasy looking, darkened cappings that may be perforated. Early in the season there may only be one or two, look carefully at the brood.

*Depressed perforated brood cells can indicate AFB*

There really is a foul smell associated with AFB. If you notice something different from a

*The match or rope test show distinct ropiness of cell contents in AFB infected cells*

*AFB affected cell. DEDJTR Victoria*

nice smell then question it!

The match test: poke a match into a suspect cell and if it pulls out "ropey" for 2-3cm, AFB is the likely suspect.

AFB is a fatal microbial disease of honey bee brood. It is a major disease of honey bees globally and is a notifiable disease in Tasmania. The spores of the bacterium can remain active and dormant for up to 80 years (Sammataro 2011). This is a significant reason for not buying second hand hive parts and why heavily infected hives have to be burnt. Robber feral bees or your bees robbing an infected hive can be a source of infection, manage your hives to prevent robbing. The beekeeper moving between hives and not cleaning equipment can also spread infection.

How does infection occur?

Larvae less than 53 hours old become infected by ingesting spores presented in their food (bee bread) but older larvae are not susceptible. The spores germinate in the larval gut into a rod like vegetative state and continue to multiply until the larva dies, usually the last 2 days of the larval stage or the first 2 days of the pupal stage, after the cell has been capped. When the vegetative form dies, it dries to a scale that lies flat on the bottom of the cell wall, it produces millions of spores with very long-term viability, making AFB a significant disease. House bees cannot remove the scale and unknowingly transfer spores that are fed to young bees.

Treatment

In Tasmania beekeepers and Government agencies have worked together to develop a best management practice guideline for American Foulbrood and European Foulbrood. Its called the The Tasmanian Foulbrood, Best Management

Practice Guideline and replaces Section 4 of the Australian Honey Bee Industry Biosecurity Code of Practice.

Essentially if AFB or EFB are suspected they are List B notifiable diseases under the Animal Health Act 1995 and an Animal Health Officer must be notified. This is also in the National Biosecurity Code of Practice.

Once a positive identification has been established at an approved laboratory a plan for disease eradication or management will be developed. This may include

- Sterilisation

- Management with the antibiotic OTC Oxytetracycline or if neither control the disease

- Destruction of the brood and all infected equipment by fire and burial with a covering of 30cm of soil

It is best to refer to the comprehensive Best Practice Guideline for treatment options beyond the precis provided.

If left unchecked AFB spreads rapidly within the brood, can spread to other hives in your apiary and to other apiaries and in the worst case kill colonies.

You must know how to recognise this highly contagious disease.

Prevention

Amateur and beginning beekeepers are encouraged not to buy used equipment, unless from a reputable beekeeper known to undertake hygienic practice and regular disease inspection.

Commercial beekeepers understand disease risks, inspect, know and manage what they are buying.

- Never allow bees access to honey, cappings, hive scrapings or dirty second hand containers.

- Use queens known to have strong hygienic traits.

- Routinely inspect the brood, especially early spring as early detection may prevent your hive from being destroyed.

- Do not move brood comb between apiaries.

- Create barrier management for any hive infected and maintain for 2 years.

European Foul Brood (EFB)

Caused by: bacterium, Melissococcus plutonius, affects brood of all ages.

What to look for: In brood comb, look for irregular brood with a mottled pattern. Infected larvae die in a coiled or twisted position and change from healthy pearly white to yellow then brown. Look in unsealed brood as larvae mostly die before cells are capped.

*Irregular brood pattern characteristic of EFB. AGRICWA*

*EFB infected cells. MV Smith*

Melissococcus plutonius is a bacterial disease of larvae of all ages. It is considered a stress disease and is most prevalent in peak brood laying periods. It is spread to larvae through infected brood food and multiplies rapidly in the gut of the larvae. Some bees survive and regurgitate the bacteria on the underside of cappings which become sources for the disease. EFB is a notifiable List B disease under the Animal Health Act 1995. If you detect or suspect EFB in your hives you must contact the government apiary specialist.

Symptoms: Larvae turn from their healthy pearly white to a yellowish colour and become unnaturally twisted. The brood pattern is irregular, with a mottled or spotty pattern of capped and uncapped cells observed when the disease reaches serious proportions.

The most obvious symptom of EFB is the colour change of larvae from pearly white to yellowish to brown and then to blackish grey.

The smell can vary from sour to foul. EFB scale is dry and loose and can be removed from cells. The match test, stick a match into a cell and pulling it back out does not produce any 'ropiness' like AFB.

Treatment

As with AFB refer to the Tasmanian Foulbrood Best Management Practice Guideline.

Notify the Apiary Specialist if you suspect you have EFB.

It can sometimes be overcome by re-queening, giving the colony a more prolific queen and allowing time for the house bees to clean the cells of diseased larvae.

Antibiotics may be able to treat mild infestations.

Significant infestation may require the destruction of the hive.

Prevention

- Prevent stress!

- Ensure adequate food resources are available, especially pollen.

- Use young queens with hygienic traits.

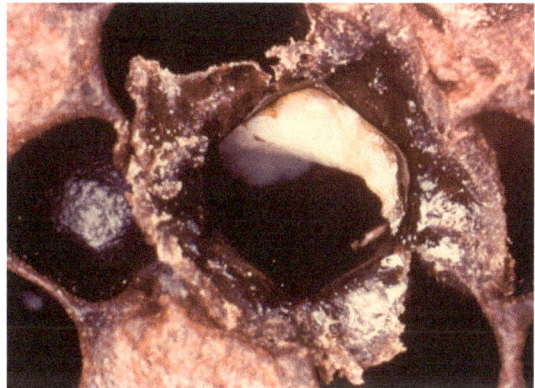

*Dead larvae caused by EFB MV Smith*

- Ensure hive is in a sheltered sunny position.

- Often disappears in a good nectar flow.

Chalkbrood

Caused by the fungus Ascosphaera apis.

What to look for: White mummified larvae first in the cells with fluffy white mycelial growth, then on the bottom board and entrance, as small hard hexagonal chalk like substance. Irregular brood pattern and perforated cappings, but this can be confused with both AFB and EFB.

*Chalk brood MV Smith*

Chalkbrood is a larval disease spread by the fungus Ascosphaera apis, affecting unsealed and sealed brood. A drop in brood temperature can trigger this disease especially when there are not enough bees to cover the brood and keep it warm. This is most likely to occur in times of rapid hive growth in spring and early summer, when there can be large temperature fluctuations. The first affected larvae are those at the outer edge of the brood.

Small startup colonies with low bee numbers are at the greatest risk of becoming chilled because they have the lowest capacity for heating and relatively large surface areas.

Chalkbrood does not usually kill the colony, but will reduce colony size and honey production. In commercial operations it can become a serious cause of loss and reduced productivity. Spores are resistant to degradation and remain viable for up to 15 years.

## Symptoms

Larvae usually die of chalkbrood after their cells have been capped.

Infection is more commonly found in worker and drone larvae than in queen larvae.

Small perforations in otherwise normal cell cappings.

Dead larvae in uncapped cells are initially fluffy white, swollen and sponge-like, covered with fungal mycelial growth and may take on the hexagonal shape of the cell. Later they become chalk like, hard hexagonal objects that are removed from cells and litter the bottom board and hive entrance. The colour can change to grey or black due to the production of fungal fruiting bodies.

## Treatment

- Replace all infected comb and remove mummified larvae from the bottom board. This removes the main sources of infection.

- For severe cases, destroy affected combs by burning.

- Re-queen with hygienic traits. Hygienic bees uncap and remove diseased brood.

- Add young adult bees and hatching brood.

- Feed sugar syrup to help hive recover.

## Prevention

- Reduce the stress on the hive, do not move too much.

- Avoid opening the hive in cold weather.

- Ensure hive site is sunny, dry and sheltered from cold winds, with good food sources available.

- Ensure the bottom board remains dry throughout the year. Make sure the hive is tipped slightly forward, supers are well maintained and the cover is water-proof and in good condition.

*Regularly re-queen with a hygienic strain, younger vigorous queens help maintain hive strength.*

- Regularly re-queen with a hygienic strain, younger vigorous queens help maintain hive strength.

- Minimise brood chamber size for over-wintering, do not over super. Four ideal supers are sufficient for brood and stores.

- Improve hive ventilation.

- New comb can reduce the incidence of chalkbrood, regularly replace brood comb on a 3-4 year rotation or when heavily discoloured.

Sacbrood virus

Caused by: Morator aetatulae, a virus.

What to look for? In open cells the larva's head changes colour becoming dark brown to black. The larvae die in an upright position in the cell.

*Sacbrood AGRIC*

If you see evidence of this disease it has probably got a foothold as worker bees are usually able to keep the hive clear of infected brood.

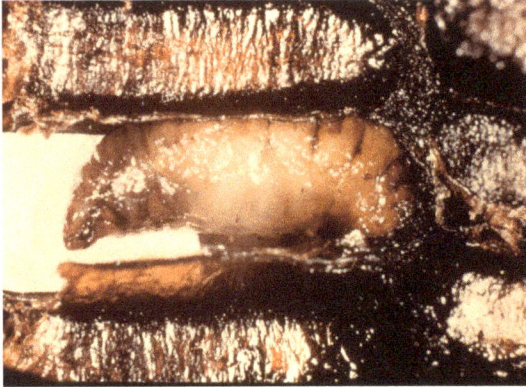

*Sacbrood dissection showing discolouration and hardening of cuticles near the head MV Smith*

The 'sac' like appearance is due to the larva's inability to shed its last skin before pupating, trapping the larvae within the 'sac' where it eventually dies. SBV can remain viable in dead larvae, honey or pollen for up to four weeks. The larvae eventually dry to a scale that resembles a gondola or a Chinese slipper, the scale is loose and easily removed, unlike AFB.

Nurse bees transmit the virus from cell to cell and robber bees from taking contaminated honey from hive to hive.

Symptoms: Perforated, sunken and or dark cell cappings are easily confused with AFB and EFB. It becomes obvious when the cell contents are withdrawn, a hanging fluid filled sac.

Treatment: There is no known chemical treatment, but in Tasmania it usually cleans up with time.

- Re-queening with hygienic traits.

- Hives often recover when the honey flow starts as there are less workers in the hive to transmit the virus.

Prevention

- Maintain strong colonies.

Pests and predators.

Braula Fly.

Caused by (Braula coeca): a small wingless fly, often incorrectly called a bee louse.

What to look for? Braula ceoca, Braula fly is a small (0.9mm wide x 1.5mm long, about the size of a flea) and it is a shiny red-brown wingless fly covered in stiff spine-like hairs. It is rounded and has six legs; it is often mis-diagnosed as Varroa destructor, a more oval shape with eight legs, not found in Tasmania, or Australia.

Within Australia Braula fly is restricted to Tasmania but found widely distributed in the rest of the world. Bees that have strong hygienic traits are better at controlling this pest. It attaches itself to the thorax of adult bees, holding on tightly with a set of comb-like structures on its front legs. When hungry, the fly moves

*Braula fly is often mistaken for Varroa by beginning beekeepers*

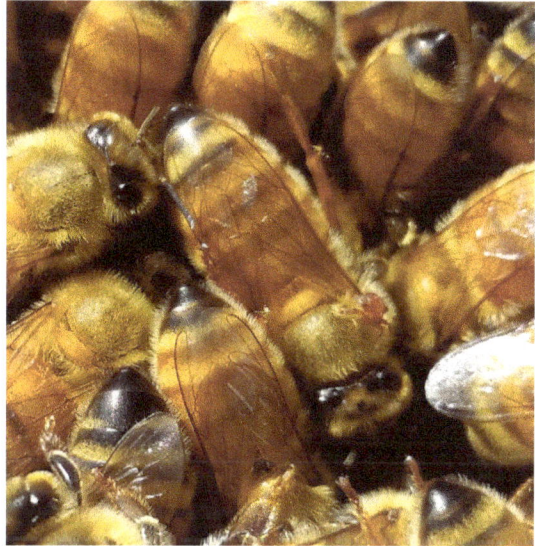

*Braula. Karla Williams*

close to the mouth parts of the bee on which it is residing, and steals some of the food fed to it by other bees. Only when a colony is broodless or small, or the queen is old is there a likelihood of finding Braula on the queen bee. It is difficult to dislodge and moves rapidly crabwise around the head and shoulders when disturbed. It is not considered a major pest as it does not damage or parasitise any stage of the honeybee life cycle. They survive in winter by being attached to adult bees and are not known to survive as individuals.

If queens become infested the braula fly can cause egg laying to decrease. Any heavily infected queen should be caught and 'deloused' immediately if she is seen to be infested. It lays small eggs (0.84mm x 0.42mm) throughout the hive but only eggs deposited on capped honey will hatch. Eggs hatch in 2-7 days depending on the temperature. As it hatches the larvae tunnel under the wax cappings, feeding on honey, creating narrow tracks about 1mm wide across the comb, giving the comb a fractured appearance, a problem for producers of comb honey. It does not affect liquid honey production as the wax cappings are removed.

Braula fly is spread through swarming and drifting honey bees. It can also be spread through the interchange of hive components from hive to hive and apiary to

*Braula fly, Braula ceoca*

apiary. It can also be spread by the removal and transport of infected honey comb.

Treatment

Regularly check your queens and catch and delouse infected queens.

• The use of small amounts of tobacco in the smoker is an effective control measure. Regular and prolonged treatment with tobacco is not recommended.

• Extracting honey by uncapping is an effective way of controlling the larval stage.

Prevention

• Regular re-queening with queens with hygienic traits.

Wax moths:

There are two species present in Tasmania, the greater wax moth Galleria mellonella, and the lesser wax moth Achroia grisella.

Wax moth larvae are very destructive, destroying hive wood ware and wax comb.

What to look for? Once you see wax moth damage you have a problem. The Galleria mellonella female is grey brown and approximately 1.3-1.9cm long and holds her wings over her body in a tent-like position. Infestation is more likely in warmer, dark environments. Larvae are white grubs with a dark end up to 2.5cm long.

*Wax moth infestation mess. David Barton Photography*

Only the larger, greater wax moth is of economic importance in Tasmania. Both are pests of stored wax and comb unattended by bees in weaker or failing colonies. Strong colonies tend to keep the moth larvae under control.

The greater wax moth appears to be more of a problem in the north of the state, causing significant damage and loss to stored comb and unused comb in the hive. Cold storage or outside storage in cooler conditions deters adult moth egg laying. They also shy away from light, so don't cover your outside stored

*Wax moth larvae. David Barton Photography*

*Wax moth larvae and small hive beetle. Kathy Keatley Garvey*

frames with a tarpaulin, find a way of keeping them dry, well aerated and open to daylight. Freezing the comb destroys all stages of the wax moth. Clean comb and foundation is not usually infested by larvae as they need the protein provided by the waste in cells from discarded cocoons.

The adult female lays eggs in any crack or suitable space in the hive or stored comb. Larvae hatch into white grubs with a brown end and can grow to 2.5cm in 18 days or longer depending on temperature. The larvae feed on comb under a protective webbing, forming tunnels across the comb and can completely destroy comb. They also tend to chew boat shaped holes in hive woodware, spinning cocoons and pupating, this can weaken and destroy frames.

In some parts of the US wax moths are considered a commodity and are bred by beekeepers for their larvae, waxworms, are used as fish bait and live pet food.

Fumitoxin® Aluminium phosphide tablets are the only product registered for use against both Greater and Lesser Wax moths. The chemical label and the permit are legal documents and the use of this highly toxic fumigant must be in accordance with the directions on the label. People intending to purchase and use Fumitoxin must have successfully completed the ChemCert course in safe chemical handling.

Prevention

- Maintain strong colonies, regularly replacing queens, strong colonies keep moth larvae under control.

- Extract honey quickly and store supers properly.

- Store empty frames or complete supers in a cool dry environment.

- Freeze frames with drawn comb to -7°C for 5 hours or -12°C for 3 hours.

(Sammataro 2011).

- Wet storage of 'stickies' also helps prevent infestation as the moths do not like moving about on sticky comb.

- Hygiene is important, keep the bottom board clean of any debris.

- Burn any heavily infected frames.

Wax moths do have a natural function as they clean up diseased feral and unoccupied hive comb preventing further infestation and providing a new clean nesting site.

### General warning

Many chemicals have been suggested and occasionally used in an attempt to control wax moth in stored combs. Commercial preparations available in supermarkets for general moth control are not suitable for wax moth control. Chemical residues have been found in honey and beeswax as a result of the use of such preparations.

Do not use products not registered for the control of wax moth (Somerville 2007).

### Other issues

Acarapis dorsalis: This parasite, an external blood sucking mite was first detected in Tasmania in 1964. It can only be seen with a microscope and lives in groups of two and three in the groove across the bee's thorax between its wings. The mite is not known to cause any loss or damage to honeybees or colonies.

Birds: Birds are usually not an issue, only the masked Wood Swallow or Martin (Artamus personatus) is a significant threat. The Wood Swallow is not on the Tasmanian Threatened Species list www.dpipwe.tas.gov.au and is evaluated as a species of least concern on the IUCN Red List.

Other predators: In certain localities, ants and spiders cause some inconvenience and loss to hives but if hygienic methods are used and a close watch kept, suitable precautions can be taken. Cockchafers have been a problem in blue gum dominated grassy woodland in the southeast (Hoskinson 2014).

*The masked wood swallow or tree martin is a bee predator*

Human issues

Chemical sprays: These are applied to orchards, crops and pastures. Bees are killed either by direct contact with them or indirectly by collecting poisoned pollen or nectar which they may well store in the hive. The second scenario also kills the larvae. Spraying is a real problem and full cooperation between fruit growers, farmers and beekeepers is essential. (Connelly 2012) noted there are 349 chemicals listed as being toxic to bees, but the list could be much greater as no labels on herbicides, fungicides or adjuvants contain bee warnings.

Vandals: When bees are migrated they are more often than not left unattended. Vandals frequently push the hives over or otherwise damage them.

Poor beekeepers: These can often be summed up in a single word – negligent. If the provision of food is neglected, the bees soon die. Poor hygiene transmits disease between hives and apiaries. At best, neglected bees give a poor financial return.

Miscellaneous and natural causes

Starvation: This is the most frequent cause of loss in Tasmania and can occur at any time from March to October or even November if the season is a bad one. It is prevalent when there is a good early spring, when populations build rapidly, and then the weather breaks up for weeks. With a larger and building population what appeared like a surplus in mid-November may all be consumed by early December if there is a run of poor weather. In the early season do not be tempted to rob too much too soon, you may be feeding your bees at the best or lose them to starvation or swarming.

Unsuitable weather, shortage of flowers and breeding on stores are all contributory causes to starvation. Beekeepers should have plenty of frames of sealed stores on hand for such an emergency or introduce artificial feeding on sugar syrup. Ensuring enough stores going into winter is a key to survival under most circumstances. In Tasmania, the late flowering of leatherwood provides excellent stores going into winter, don't be too greedy!

Pollen infected with certain moulds: If fed to larvae, mouldy pollen has been known to cause death either directly or indirectly by invasion after the cell has been sealed over. This is not a serious cause of loss and can be

*A hive starved in the summer dearth.*

readily distinguished from the symptoms of AFB.

Weather: The weather in Tasmanian springs can vary within a day from the low 20's to almost freezing with wintery blasts. Bees can be caught out on a flight and be killed by such cold and or wet snaps.

Ensuring your hive is weather proof is an essential practice throughout Tasmania, but especially in higher elevations and wetter areas. As the brood nest continually expands and contracts, large areas of chilled brood can occur in early swarms and weaker hives where there are not enough adult bees to cover the brood. A cold night causes the cluster to contract so that the outer brood is left uncovered and dies of cold. The cover should be insulated with a felt or similar cover over the top frames and the size of the entrance reduced in continual cold weather.

Overheating is not common in Tasmania, but adequate and close supplies of clean water is very important.

Potential Threats

The small hive beetle

(Aethina tumida Murray) is not found in Tasmania or the Northern Territory, but all other States have reported it. Quarantine measures are in place to prevent its introduction. Importation to Tasmania of used hive elements is banned. It is included as a precautionary bio-security measure.

In its native sub-Saharan Africa, it is considered a secondary apiary pest. However, where SHB incursions have occurred and European honey bees are managed in hot humid locations e.g. south-eastern USA 1998 and eastern Australia 2002, SHB have become a major pest of beekeeping industries (Annand 2008).

*Small hive beetle is not found in Tasmania. Beekeepers must remain vigilant*

What to look for? Aethina tumida is a small beetle (5-7mm long x 3-5mm wide) Adults have a club antennae, a shield shaped thorax and a wing case shorter than the tail. Eggs are about 2/3 the size of bee eggs in clusters away from access by worker bees. Larvae begin very small, about 1mm and grow to 10-15mm. They eat through comb, slime honey and fermented liquid can leak from cells. The smell of rotten oranges will be obvious.

Both adult beetles and larvae infest hives where they feed on honey and pollen. Larvae cause the most damage by burrowing into comb, eating brood, honey and pollen. The larvae also carry a yeast Kodamaea ohmeri that contaminates honey and causes it to ferment and give off an odour like sour or rotting oranges. Heavy infestations cause the hive to be 'slimed' out and may cause the hive to die or the bees to leave. Similarly, stored supers of honey such as those in honey extracting plants or extracted combs can also be ruined by infestation by adult beetles and larvae.

The small hive beetle may be spread between colonies and apiaries when bees, package bees, used hive components and hives are moved during normal management practices. Hive components may provide places for adult beetles to hide

Symptons:

Field signs are numerous larvae burrowing and tunnelling in brood and honeycomb cells. Adult beetles may not always be easily observed. Honey may weep from comb where cell walls and caps have been pierced by the activities of the larvae. The colour of honey may darken due to contamination of faeces deposited by the larvae. Fermentation and frothing of honey in hives or in combs of honey stored in honey houses with an odour similar to decaying oranges are typical signs of infestation. Fermented honey may drip and collect on the hive bottom board or the honey house floor. Trails of fermented honey will be left on the honey house floor by larvae that crawl from stored honey supers as they try to reach soil to pupate.

*Small hive beetle larvae cause significant damage*

When a hive is opened, the adult beetles usually run quickly from the combs away from light to dark parts of the hive particularly towards the back. The rear part of the bottom board is a favourite hiding place. Larvae of the small hive beetle may be found in combs containing brood and/or honey, but they prefer combs that contain pollen.

Prevention

- Do not move hive components to uninfected apiaries.

- Strong colonies are one of the best defence against strong infestation.

- Regularly re-queen with queens that have hygienic traits and known SHB reduction.

- Beekeepers should maintain good hygiene practices of removing burr comb, brace comb taking it off site and burn or store it in a covered container, remove debris from bottom boards.

- Cool rooms kept at 15°C or below for storage of supers comb prevent adult SHB laying eggs.

- Maintain supers, repairing cracks or crevices.

- Extract honey as soon as possible after robbing.

Health Caution: SHB larvae carry the yeast Kodamaea ohmeri that poses a threat to immuno-compromised people. Handle infected hive components using appropriate precautions.

Varroa mite

Caused by : Varroa destructor, commonly known just as Varroa, is not in Australia but in our near neighbour New Zealand and Indonesia as well as the majority of the world. A new mite related to V. destructor, V. jacobsoni was discovered in Papua and New Guinea.

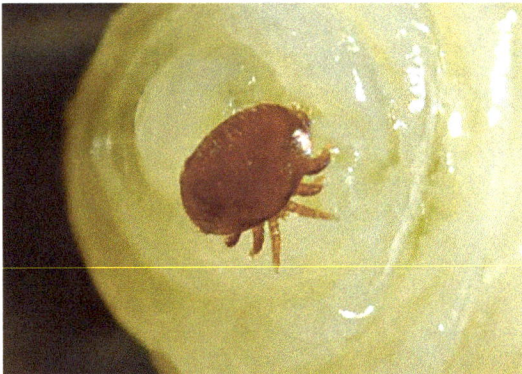

*The dreaded Varroa destructor, more commonly called Varroa or Varroa mite. Kathy Keatley Garvey*

*Varroa mites in drone cells. Kathy Keatley Garvey*

Varroa mites are considered to be the most serious threat to honeybees globally. This sesame seed sized creature has the ability, if not treated, to destroy the hive, sucking the bees haemolyph (blood) and introducing other destructive viruses. The adult Varroa is about the size of our Braula, but more oval than round, similar in color, and can often be seen scuttling around the thorax of the bee or hiding between the segments of the thorax, the obvious difference is the Braula, a fly has six legs and Varroa, a mite has eight.

Varroa mite is a deadly parasite of the European honey bee which has spread to all inhabited continents except Australia. In the US and Europe, Varroa kills 95–100 per cent of unmanaged hives within three to four years of infestation (DAFF 2011). Its introduction around the world has caused devastation to apiary industries; it has wiped out wild or feral populations and caused huge losses to all beekeepers. It has been associated with the phenomenon of Colony Collapse Disorder CCD in the US and Europe. Kim Flottum, US bee author emphasises the importance of understanding how to manage for and treat this destructive creature, "I liken controlling varroa mites in beekeeping as important as knowing which side of the road to stay on when learning to drive. If you don't control varroa, your bees die, period". While it is not in Australia or Tasmania, it remains a significant national threat. The following provides some detail to understand and identify this mite.

The damage varroa does to bees is subtle and not clearly understood, but is probably responsible for;

- Reduced flight activity for foraging bees.

- Weight loss (6-25%).

- Reduced life span (by 34-68%).

- Reduced blood volume (by 15-50%) when parasitized by mites.

- External damage (chewed wings, legs, stunted growth) if more than five minutes in one cell.

- Transmission of virus and other pathogens.

*Varroa on a forager. Kathy Keatley Garvey*

(Sammataro & Avitable 2011).

Symptoms

The suite of symptoms from Varroa are called varroosis. If colonies have high levels of varroa mites the following may be observed. Taken from (Sammataro and Avitable 2011)

- Disfigured stunted adult bees with deformed wings, legs or both and can be seen crawling on the ground.

- Bees seen discarding infested or deformed larvae and pupae.

- Pale or dark reddish brown spots on otherwise white pupae.

- Spotty brood pattern and the presence of diseases.

- Dead colonies in the late summer, straight after honey harvest.

- Queen supersedure more than normal.

- Foulbrood and sacbrood symptoms present.

- AFB symptoms existing, but no ropiness, odor or brittle scale present.

- No predominant bacterial disease found.

Varroa mite latches onto adult bees and feeds by piercing the body of the bee and sucking the haemolymph. Bees can survive with Varroa, but the introduction of viruses such as deformed wing virus further threatens their existence.

The mite life cycle follows that of the bee. The only mites seen in the hive are adult females. The female mites are attracted to drone brood pheromone. When the bee larvae are 7-8 days old the mite moves into the cell and hides at the bottom under food fed to the larvae. Here she avoids detection from nurse bees and avoids being trapped in the silk of the larva. When the cocoon is finished the mite pierces a hole in the pupa and lays her first egg within 30 hours, a sterile egg that becomes a male, followed every 30 hours by fertile eggs that become females. Feeding from the pupa continues, then the male after about six days mates with one or two of his sisters and dies in the cell. When the weakened bee completes metamorphosis and emerges, the mother and daughter mites crawl out well fed and the vicious cycle continues.

There is a wealth of information available on identifying and detecting varroa mites, see Bibliography.

Biosecurity threats to the honey bee industry

Australia is currently free from some of the most significant pests of honey bees that occur elsewhere in the world, notably the Varroa mite (Varroa destructor and V. jacobsoni), Tropilaelaps mite (Tropilaelaps clareae and T. mercedesae) and Tracheal mite (Acarapis woodi).

The establishment of these bee pests in Australia would be challenging for the honey bee industry, causing losses of production in both honey bee products and pollination services. Every other beekeeping region in the world has experienced large reductions in the number of beekeepers and number of hives after the introduction of Varroa mite. In addition to this, feral honey bee colonies that previously provided free pollination for crops all but disappeared.

Established bee pests also cause significant economic and social harm and need strategic management to limit the impact to individual beekeepers and the broader industry and economy. In particular, American foulbrood (AFB) (Paenibacillus larvae) is present in all Australian states and territories and is the most fatal and costly established pest. Evidence shows that current policies and systems to manage pests such as AFB are ineffective with pest problems worsening. Other established pests, such as Small hive beetle (Aethina tumida) and Nosemosis (Nosema sp.) cause ongoing challenges for beekeepers.

Remember

- Prevention is better than cure.

- Look for hygienic traits in your queens.

- Regularly requeen.

- Systematically replace discoloured brood comb.

- DO NOT USE SECOND HAND EQUIPMENT (spores can remain dormant for decades).

- Use hygienic practices yourself, clean equipment, wash protective clothing.

- Do not lift the lid in wet cold conditions!

- Inform yourself, carefully observe, be vigilant.

Bee Biosecurity Manual

Plant Health Australia, the Australian Honey Industry Council AHBIC, The RuraL Industries Research and Development Coproration, the The Wheen Bee Foundation and Horticulture Innovation Australia have developed the Bee Biosecurity Manual 2016 to educate beekeepers in practical biosecurity measures.

*Burning old equipment helps prevent disease spread.*
*CAUTION: wax is highly flammable*

NATIONAL
**BEE**
**BIOSECURITY**
PROGRAM

# CHAPTER SEVENTEEN

This is a free download from www.beeaware.org.au

Honey bee biosecurity is a set of measures designed to protect your honey bees from the entry and spread of pests. Honey bee biosecurity is the responsibility of every beekeeper and every person visiting or working in an apiary.

Implementing honey bee biosecurity is essential for your business. If an exotic or endemic pest establishes in an apiary, business costs will increase (for monitoring, hive management, additional chemical use and labour), productivity will decrease (yield and/or colony performance) and markets may be lost. The health of the honey bee industry also ensures the continued success of many other plant industries that rely on honey bees for pollination.

Early detection and immediate reporting increases the chance of an effective and efficient eradication.

1. Be aware of biosecurity threats

2. Use pest-free honey bee stock and apiary equipment

3. Keep it clean

4. Check your apiary

5. Abide by the law

6. Report anything unusual

If you suspect a new pest – report it immediately to the

Exotic Plant Pest Hotline. 1800 084 881

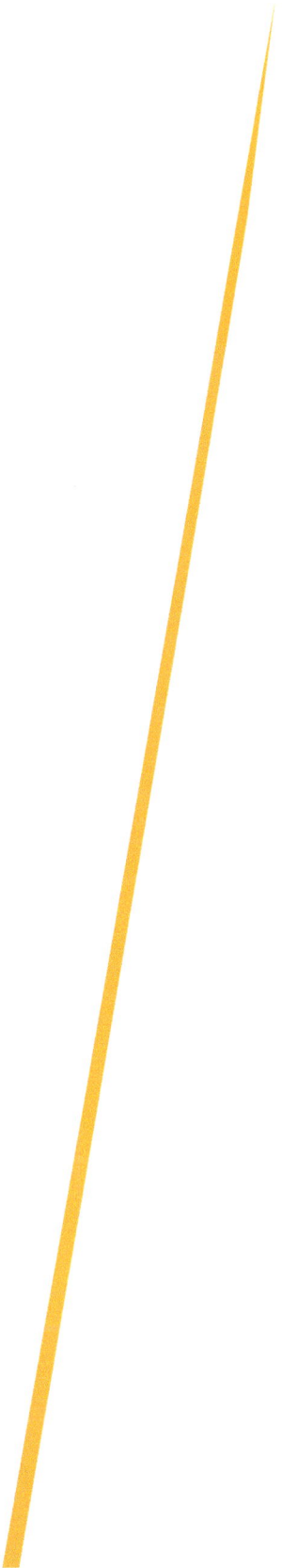

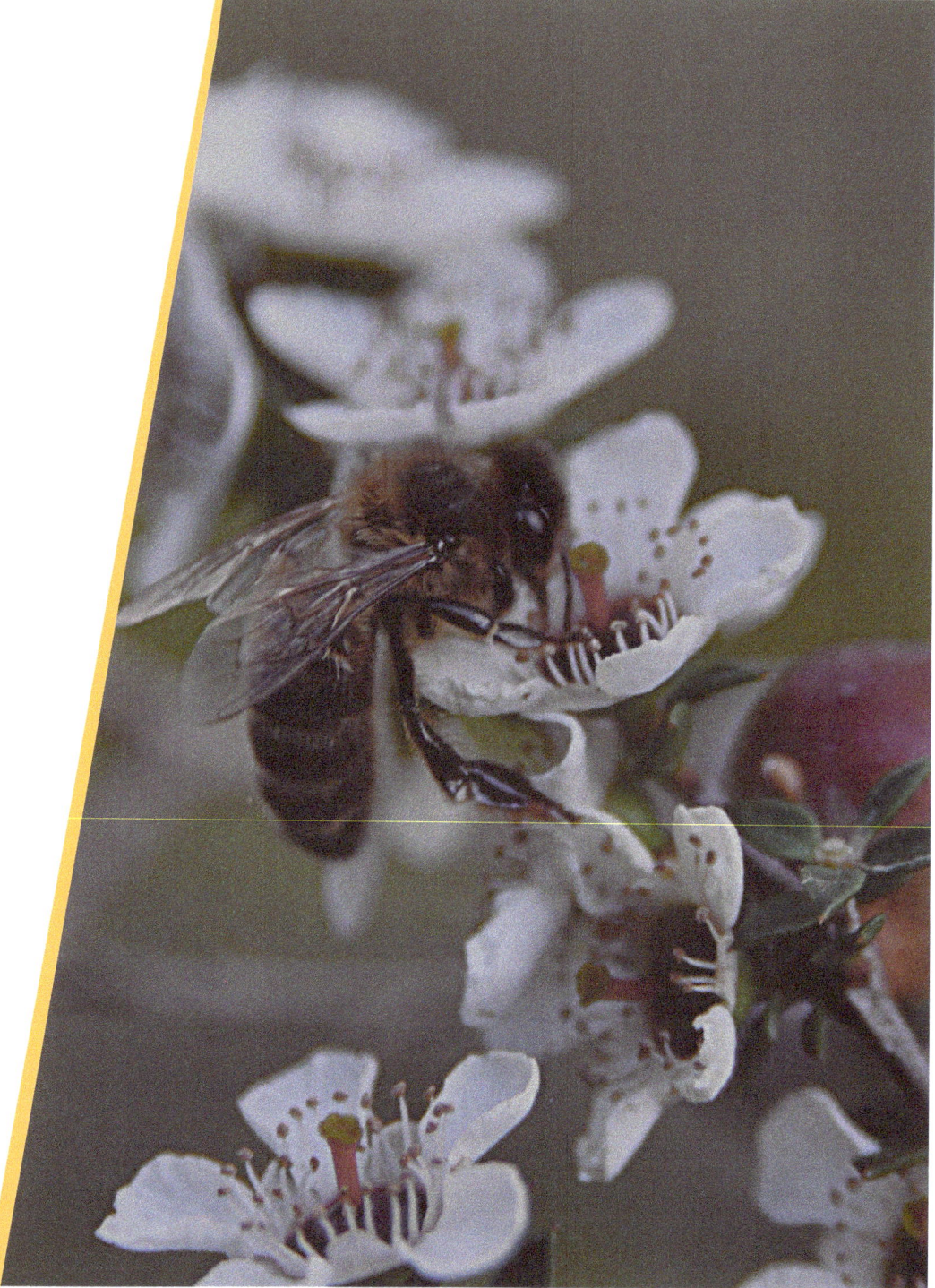

# CHAPTER EIGHTEEN

Some people prefer to explore what may be considered a 'close to nature' approach to beekeeping where the inside environment of the hive resembles the comb structure of a wild hive. A number of examples exist and historically prior to the Langstroth development this was the main means of beekeeping. Images of straw skeps, hollow logs and other devices are in the historic record and are still being used in some countries.

Top bar hives provide this close to nature approach. The most common examples are the Kenyan Top Bar Hive and the Warre Hive. The Kenyan top bar hive was developed for ease of construction from available materials in the developing world. The Warre hive was developed by a French abbot, Abbé Emile Warré to provide an easily constructed peasant's hive which in fact mimicked a tree hollow, the natural home for bees in the wild.

Bisosecurity Warning!

All hives used for beekeeping in Australia must have removable comb that can be inspected for disease. Alternatives to Langstroth hives must make provision for frames to be removable. The two top-bar hives described achieve this by having either a strip of "starter' foundation attached to the underside of the top-bar or roughened bars for comb guides.

A closer to nature approach is based on the top bar hives where only starter foundation is provided and the bees build vertical comb as in a natural state. The Kenyan top bar hive is a single tiered horizontal box with sloping sides and frameless top bars, developed for ease of construction and management in the developing nations. The Warre' hive developed by the French monk Abbe Emile Warre' in the early 1900's provides an internal environment like the inside of a tree hollow, and is based on stackable supers and foundation-less frames.

*Australian regulations require frames of comb to be removeable for inspection, this is an important biosecurity measure to inspect for disease and general colony health.*

Australian regulations require frames of comb to be removeable for inspection, this is an important biosecurity measure to inspect for disease and general colony health. Where removable frames are not used, natural beekeepers are able to dismantle combs to support these biosecurity measures.

*A Kenyan top bar hive can be made from scrap material*

Top Bar

The Kenyan top bar hive is a rectangular horizontal box with sloping sides, a roof of some description, top bars and a leaderboard to control expansion of the colony. The top bars can be shaped with a down pointing V and most often will have a strip of starter wax.

*They can be built from scrap material such as untreated pallet wood or more durable material.*

They can be built from scrap material such as untreated pallet wood or more durable material. As the top bars are 'frameless', the comb is simply cut from the bars, crushed and strained. This low cost minimal equipment approach is ideal for backyard beekeeping. It must be noted that the basic principles of beekeeping are the same, but the management of a top bar hive requires some experience. The configuration may have to be adjusted for different climates.

There are groups of beekeepers passionate about this as a lower impact alternative

*A Warré hive in a garden setting. NaturalbeekeepingTasmania*

and they can provide support including courses specifically developed for top bar beekeeping. There are also a number of active blog sites and pages that can be helpful to anyone wanting to find out more.

Warré Hive

The Warré hive is a close to nature approach to beekeeping refined by the French abbot Emile Warré in the early 1900's. The main concept is a square stackable series of boxes that provide an environment that protects the heat and scents of the brood nest.

Warré discovered that 210mm is a crucial height for the boxes to enable them to be readily separated. The square narrow tall form of the hives emulates the natural comb forming of a swarm and the way comb is constructed in tree hollows. There are a number of Warré specialist beekeepers in Australia listed in the bibliography.

*Comb construction in a Warre*

The  Warré hive is a tiered top-bar hive comprising a stack of at least two boxes each

*Modified frames used in Warre hives by Natural Beekeeping Tasmania provide the ability to remove frames for inspection. NaturalbeekeeingTasmania*

*Wood shavings used as the insulating quilt Natural Beekeeping Tasmania*

of internal dimensions 300 x 300 x 210 (deep) mm with eight 8 x 24 mm top-bars at 36 mm centres. The floor, a plain board, is notched to form a 120 mm wide entrance and has an alighting board nailed underneath (Heaf 2008).

Comb construction begins in the top box and continues in subsequent boxes below. The main issue with Warré hives is that additional boxes are from below. During a heavy honey flow this can require two people or a specialised lifting device.

The hive has a number of key elements to provide maintenance of temperature, the natural scents of the hive, comb form and venting.

Innovation

Flow Hive:

Many people; beekeepers and non-beekeepers alike watched in 2015 as a global phenomenon rapidly grew through a crowd funding site, The Flow™ Hive emerged on the global scene. The worlds' response shocked its developers who were simply seeking a modest level of development support and took the world of beekeeping by storm.

The Flow™ Hive provides an innovative approach to on site honey collection using a modified comb system manufactured from plastic. Beekeeping practices remain essentially the same, this is not a turn the key and get honey system, it requires attention to good beekeeping to provide the on-site honey collection.

The system uses a 'split' comb approach (go to www.flow.com site for details).

The mainstream industry was quick to criticise the approach as it was such a rapid influence on people with no beekeeping background and could lead to poor beekeeping and possible disease issues. While this may be true, it is imperative that beekeepers and beekeeping organisations around the world embrace those non-beekeepers who have purchased a Flow™ hive and engage with them. The momentum generated by the Flow™ Hive should be viewed as a positive, gaining the attention and affection of the many people who have a great interest in honeybees.

The sales pitch has been, Turn the tap and watch as pure, fresh, clean honey flows right out of the hive and into your jar. No mess, no fuss, no expensive processing equipment and without disturbing the bees. Many established beekeepers were concerned that this type of message missed the need to manage honeybees. However, the developers of the Flow™ Hive, Cedar and Stuart Anderson of Byron Bay NSW emphasise the importance of proper beekeeping and the need for non-beekeepers purchasing a Flow™ Hive to inform themselves, join a beekeeping club and do a beekeeping course.

Close to nature beekeepers don't agree with the use of plastic in beekeeping, this applies equally to commercial beekeeping and the use of plastic hive components including foundation, a practice that has been used for many years. Needless to say many close to nature beekeepers are strong antagonists of the Flow™ Hive.

*The success of the honey capturing system will depend on the strength of the honey flow, the type of honey and especially honey that crystallises rapidly.*

The success of the honey capturing system will depend on the strength of the honey flow, the type of honey and especially honey that crystallises rapidly.

It is also fundamental that the tap is not turned on too soon. The cells must be at least 75% capped before cracking the comb system allowing the honey to flow. The principle is the same as applied to uncapping and extracting regular capped comb. Honey must be at approximately 18% moisture content when extracted to prevent fermentation. The Flow™ Hive has observation windows where filled and capped comb can be observed.

The system works when there is a healthy population of well managed bees and a nectar flow. Once the comb is full of honey the split comb mechanism system is cracked by turning the crank. This may be quite difficult but can be undertaken in stages then the honey flow is almost instantaneous.

While practical application of the Flow™ Hive is still in its early adoption with only one full season around the world, lessons have been learnt and there are many success stories. It must be emphasised that this is a honey collecting system that requires good beekeeping practices to

*While practical application of the Flow™ Hive is still in its early adoption with only one full season around the world*

manage a healthy strong colony of bees as with any beekeeping initiative. All the same precautions are required for personal safety and bee health. It is not a closer to nature approach, but may not disturb the bees during extraction.

The developers, Cedar and Stuart Anderson are to be applauded for bringing to market an innovative approach to honey collection and a system that has fanned the flame of global interest in honeybees. The challenge is for existing beekeepers to embrace these beginnings and help this throng of new beekeepers become good stewards of their bees.

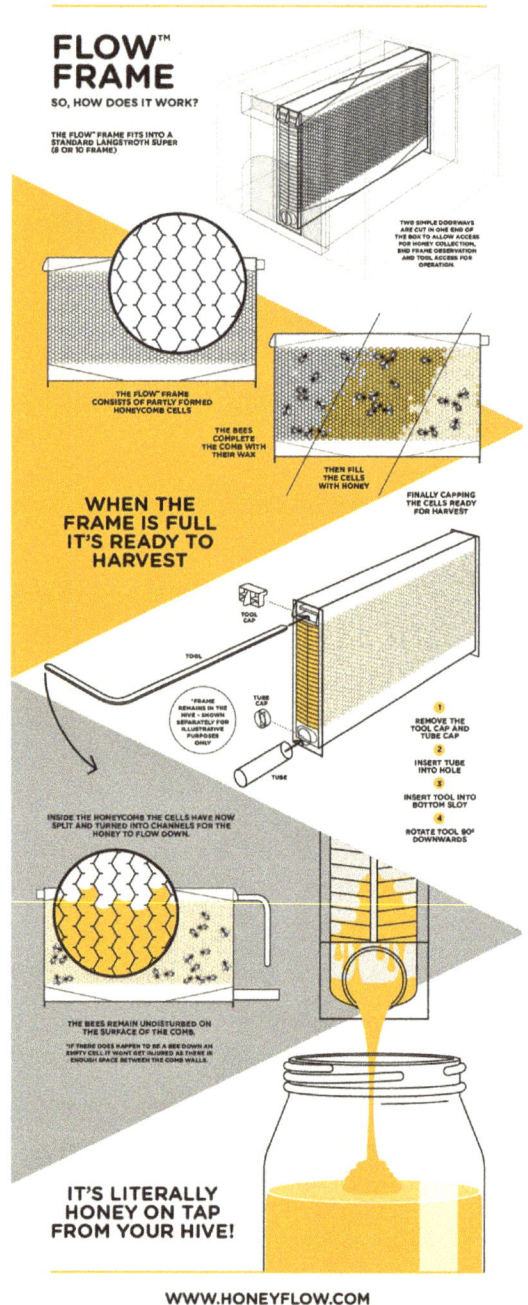

**FLOW™ FRAME**

SO, HOW DOES IT WORK?

THE FLOW™ FRAME FITS INTO A STANDARD LANGSTROTH SUPER (8 OR 10 FRAME)

TWO SIMPLE DOORWAYS ARE CUT IN ONE END OF THE BOX TO ALLOW ACCESS FOR HONEY COLLECTION, END FRAME OBSERVATION AND TOOL ACCESS FOR OPERATION

THE FLOW™ FRAME CONSISTS OF PARTLY FORMED HONEYCOMB CELLS

THE BEES COMPLETE THE COMB WITH THEIR WAX

THEN FILL THE CELLS WITH HONEY

FINALLY CAPPING THE CELLS READY FOR HARVEST

**WHEN THE FRAME IS FULL IT'S READY TO HARVEST**

TOOL CAP

TOOL

TUBE CAP

*FRAME REMAINS IN THE HIVE - SHOWN SEPARATELY FOR ILLUSTRATIVE PURPOSES ONLY

TUBE

1 REMOVE THE TOOL CAP AND TUBE CAP
2 INSERT TUBE INTO HOLE
3 INSERT TOOL INTO BOTTOM SLOT
4 ROTATE TOOL 90° DOWNWARDS

INSIDE THE HONEYCOMB THE CELLS HAVE NOW SPLIT AND TURNED INTO CHANNELS FOR THE HONEY TO FLOW DOWN

THE BEES REMAIN UNDISTURBED ON THE SURFACE OF THE COMB

*IF THERE DOES HAPPEN TO BE A BEE DOWN AN EMPTY CELL IT WONT GET INJURED AS THERE IS ENOUGH SPACE BETWEEN THE COMB WALLS.

**IT'S LITERALLY HONEY ON TAP FROM YOUR HIVE!**

WWW.HONEYFLOW.COM

*How it works. Flow™*

*A Flow™ frame removed for inspection. Flow™*

*A Flow™ set up and producing honey. Flow™*

Disclaimer: This chapter has been independently edited by Ronnie Voigt of Natural Beekeeping Tasmania

Adams CJ, Boult CH, Deadman BJ, Farr JM, Grainger MNC, et al. (2008) Isolation by HPLC and characterisation of the bioactive fraction of New Zealand manuka (Leptospermum scoparium) honey. Carbohydrate Research 343: 651–659.

Annand, N. 2008 Small Hive Beetle Management Options. Primefact 784, Department of Primary Industry NSW.

Arcot J, Brand-Miller J 2005 A preliminary assessment of the Glycemic Index of honey. Rural Industries Research and Development Corporation

RIRDC Publication No. 05/027

Ayton, H. (1991) Beekeeping in Tasmania. Department of Primary Industry Tasmania

Bang L.M., Bunting, C., Molan P., 2003

The Effect of Dilution on the Rate of Hydrogen Peroxide Production in Honey and Its Implications for Wound Healing

The Journal of Alternative and Complementary Medicine

Volume 9, Number 2, 2003, pp. 267–273

Barrette, E. (2010) Plant a Bee Garden. Gaiatribe. http://gaiatribe.geekuniversalis. com/2010/02/17/plant-a-bee-garden/.

Black, J. 2006 Honeybee Nutrition A review of research and practices.

Bourke, Lindsay & Yonsoon Personal Communication and tremendous support

Chittka L: 2004 Dances as Windows into Insect Perception. PLoS Biol 2/7/2004: e216.

Clarke, Whitney, Sutton & Robert, 2013 Detection and Learning of Floral Electric Fields by Bumblebees. Science http:/dx.doi.org/10.1126/science.1230883

Connelly, D. 2012 Honeybee pesticide poisoning: a risk management tool for Australian farmers and beekeepers. Rural Industries Research and Development Corporation. RIRDC Publication No. 12/043

Cowen, L. Personal Communication, Laurie has been an amazing help throughout this project, giving practical tips making the words in the book real.

Cramp D., 2008 A Practical Manual of Beekeeping How to keep bees and develop your full potential as an apiarist. How to Books, Spring Hill

Crane, E. 1990 Bees and Beekeeping: science, practice and world resources

# BIBLIOGRAPHY

Crane, E. 1999 The world history of beekeeping and honey hunting.

DAFF 2011 A honey bee and pollination industry continuity strategy should Varroa mite become established in Australia. Department of Agriculture Fisheries and Forestry 2011

D'Arcy, B. 2007 High-power Ultrasound to Control of Honey Crystallisation

Rural Industries Research and Development Corporation. Publication No. 07/145

Dawes, J. and Dall, D. 2014 Value Adding to Honey

Rural Industries Research and Development Corporation.

RIRDC Publication No. 13/123

De Groot, A.P. (1953) Protein and amino acid requirements of the honeybee Apis mellifera. Physiologica Comparata et d'Ecologia. Vol 3, pp. 195-285

Dotimas, E.M. and Hider, R.C. 1987. Honeybee venom. Bee World, 68 (2): 51-70

Ewington, P & M Personal communication, Peter and Margaret are always a source of encouragement and information

Ginat, Y. Personal communication, an inspirational beekeeper who is always so willing and helpful

Goodman, C. (2010) Bee-Friendly Landscaping. http://www.networx.com/article/bee-friendly-landscaping.

Goodwin, M. 2012 Pollination of Crops in Australia and New Zealand,

Rural Industries Research and Development Corporation RIRDC Pub 12/059

Heaf, D. (2008) Sustainable and Bee-friendly Beekeeping Australian Bee Journal, 89(2) Feb 2008, 5-12

Hewitt, I. 2014 Personal communication, Ian is a very experienced beekeeper with the label EE's Bees, always helpful

Hornitzky, M. 2007 A Study of Nosema ceranae in Honeybees in Australia

Rural Industries Research and Development Corporation.

RIRDC Publication No. 11/045

Hoskinson, H., Personal communication, Hedley is a most experienced "unretired" master beekeeper, a source of endless wisdom, always available.

IBRA (2008) Garden Plants Valuable to Bees. International Bee Research Association. Cardiff, Wales.

Irish J, Blair S, Carter DA (2011) The Antibacterial Activity of Honey Derived from Australian Flora. PLoS ONE 6(3): e18229. doi:10.1371/journal.pone.0018229

Jazz, H. (2009). Usage of Honey in Ancient Europe. Durham Archaeological Society.

Jones, P. Personal communication

Leech, M. (2012) Bee Friendly: A planting guide for European honeybees and Australian native pollinators. Rural Industries Research and Development Corporation. RIRDC Publication no. 12/014

Leech, M. (2009) A Field Guide to Native Flora Used by Honeybees in Tasmania. Rural Industries Research and Development Corporation.

RIRDC Publication no. 09/149

Leech, M. (2009) Tasmanian Floral Resources for Honeybees - Focus on tea tree Rural Industries Research and Development Corporation. RIRDC Publication no. 09/153

Lomsadze, G. 2012 Report: Georgia Unearths the World's Oldest Honey, Tamada Tales http://www.eurasianet.org/node/65204

Mavric, E., Wittman,S., Barth, G., and Henle, T. 2008 Identification and quantification of methylglyoxal as the dominant antibacterial constituent of Manuka (Leptospermum scoparium) honeys from New Zealand

Molecular Nutrition and Food Research. 2008, 1 52, 000 – 000

McGovern, P. E.; Zhang, J; Tang, J; Zhang, Z; Hall, G. R.; Moreau, R. A.; Nuñez, A; Butrym, E. D. et al. (December 6, 2004). "Fermented beverages of pre- and proto-historic China".

Proceedings of the National Academy of Sciences of the United States of America 101 (51): 17593–8.

Molan, P.C. 2006 Using honey in wound care. International Journal of clinical Aromatherapy 2006 Vol 3 Issue 2.

Molan, P.C. 2007 Honey and Medicine: Past, Present and Future
Abstracts 1st International Conference on the Medicinal Use of Honey (From Hive to Therapy) Malaysian Journal of Medical Science Jan 2007:14(1): 101-127

Mussen, Eric Persoanl communication. A stalwart of the Amrican beekeeping acadaemia and industry.

Mwipatayi BP, Angel D, Norrish J, Hamilton MJ, Scott A & Sieunarine K. The use of honey in chronic leg ulcers: a literature review. Primary Intention 2004; 12(3):

# BIBLIOGRAPHY

107-108, 110-112.

Pavel C., Mărghitaş, L., Bobiş, O., Dezmirean, D., Şapcaliu, A., Radoi3, I., Mădaş, M., 2011 Scientific Papers: Animal Science and Biotechnologies, 2011, 44 (2)

Pheromones: A list of honey bee pheromones https://en.wikipedia.org/wiki/List_of_honey_bee_pheromones

Plant Health Australia, Australian Honey Bee Industry Council, Rural Industries Reasearch and Development Corporation, Horticulture Innovation Australia, , Wheen Bee Foundation, 2016, A Biosecurity Manual for Beekeepers.

Roberts, J. Anderson, D. Wee Tek Tay 2014 Varroa jacobsoni: a new pest of European honeybees. Project Summary Rural Industries Research and Development Corporation. Publication No. 14/005

Sammataro, D. and Avitabile, A. 2011 The Beekeepers Handbook 4th Edition Comstock Publishing, Cornell University Press

Shepherd, M. 2004 General Gardening Advice for Attracting Bees and Other Pollinators, Adapted from Xerces Society Pollinator Conservation Program.

Somerville, D. 2002 Honey & pollen flora suitable for planting in south-eastern New South Wales, Agnote DAI- 115, NSW Agriculture, Sydney.

Somerville, D. 2005 Fat Bees Skinny Bees – a manual on honeybee nutrition for beekeepers. Rural Industries research and Deveopment Corporation, RIRDC Publication, 05/054

Somerville 2007 Wax Moth, Prime Facts Primefact 658 NSW Department of Primary Industry.

Somerville, D. 2013 Responsible Beekeeping. Primefact 1288 Department of Primary Industries NSW Government

Spivak M 2013 Why bees are disappearing. TED Lecture http://www.ted.com/talks/marla_spivak_why_bees_are_disappearing/transcript?language=en

Stephens Family Personal communication and a great support throughout my beekeeping research and involvement in the industry

Tasmanian Crop Pollinators Association 2012, Tasmanian Crop Pollinators Association Code of Practice. http://dpipwe.tas.gov.au/Documents/Code-of-Practice-Tas-Crop-Pollination-Aug-12.pdf

Tew, J. in Flottum, K. 2011 The Complete and Easy Guide to Beekeeping, page 135 "Most people get into beekeeping because of their curiosity about bees, but they leave beekeeping because of the nightmare of harvesting honey". Dr. James. E. Tew, Extension Specialist for Apiculture, Ohio.

Warhurst, P. and Goebel, R. 2013 The bee book. Beekeeping in Australia

Warré A. Beekeeping For All Translated from the original French version of L'Apiculture Pour Tous (12th edition)1 by Patricia and David Heaf. Sixth electronic English edition thoroughly revised February 2010.

Wolfhagen, J. Personal Communication and a great support over many years

DPI NSW 2013, Bee Agskills, A Practical Guide to Farm Skills

http://www.dermnetnz.org/treatments/honey.html

http://www.ncbi.nlm.nih.gov/pmc/articles/PMC3609166/

Honey: its medicinal property and antibacterial activity

Asian Pac J Trop Biomed. Apr 2011; 1(2): 154–160

Useful Web Sites

Tasmanian Beekeepers Association www.tasmanianbeekeepers.org.au

Southern Tasmanian Beekeepers www.southerntasmanianbeekeepers.org.au

Australian Beekeeper Journal www.theabk.com.au

Australian Apaiarist network www.aussieapiaristsonline.net

Bees for Development   http://beesfordevelopment.org/

Apimondia The International Fedration od Beekeepers Associations ww.apimondia.com/en

British Beekeepers Association. www.bbka.org.uk

International Bee Research Association www.ibrabee.org.uk

Rural Industries Research and Development Corporation RIRDC Honeybee and Pollination www.rirdc.gov.au/research-programs/animal-industries/honeybee

Tasmanian Web Sites and Facebook Pages (A more comprehensive list can be obtained by searching Google)

Blue Hills Honey www.bluehillshoney.com

Heritage Honey www.heritagehoney.com.au

Wellington Apiaries www.wellingtonapiary.com

# BIBLIOGRAPHY

Tasmanian Honey Company   http://www.tasmanianhoney.com Shop: 24A Main Road Perth Tasmania 7300

R. Stephens www.leatherwoodhoney.com.au Shop: Mole Creek Tasmanian

Miellerie Honey Tasmania  www.miellerie.com.au

Honey Tasmania http://www.honeytasmania.com/ Shop, Exeter Tasmania

The Honey Farm http://www.thehoneyfarm.com.au/ Shop: Chudleigh Tasmania


Beekeeping Supplies

Hollander Imports, Brooker Highway, Hobart Tasmania

Heritage Honey, 21 - 23 Springfield Ave, Moonah TAS 7009

Wells Home and Timber Latrobe Tasmania www.wellslatrobe.com/rural

A Google search provides the most up to date list of Australian beekeeping suppliers.

eBay: buyer beware, there are some good buys. Shop local if you can, if you don't use it you lose it.

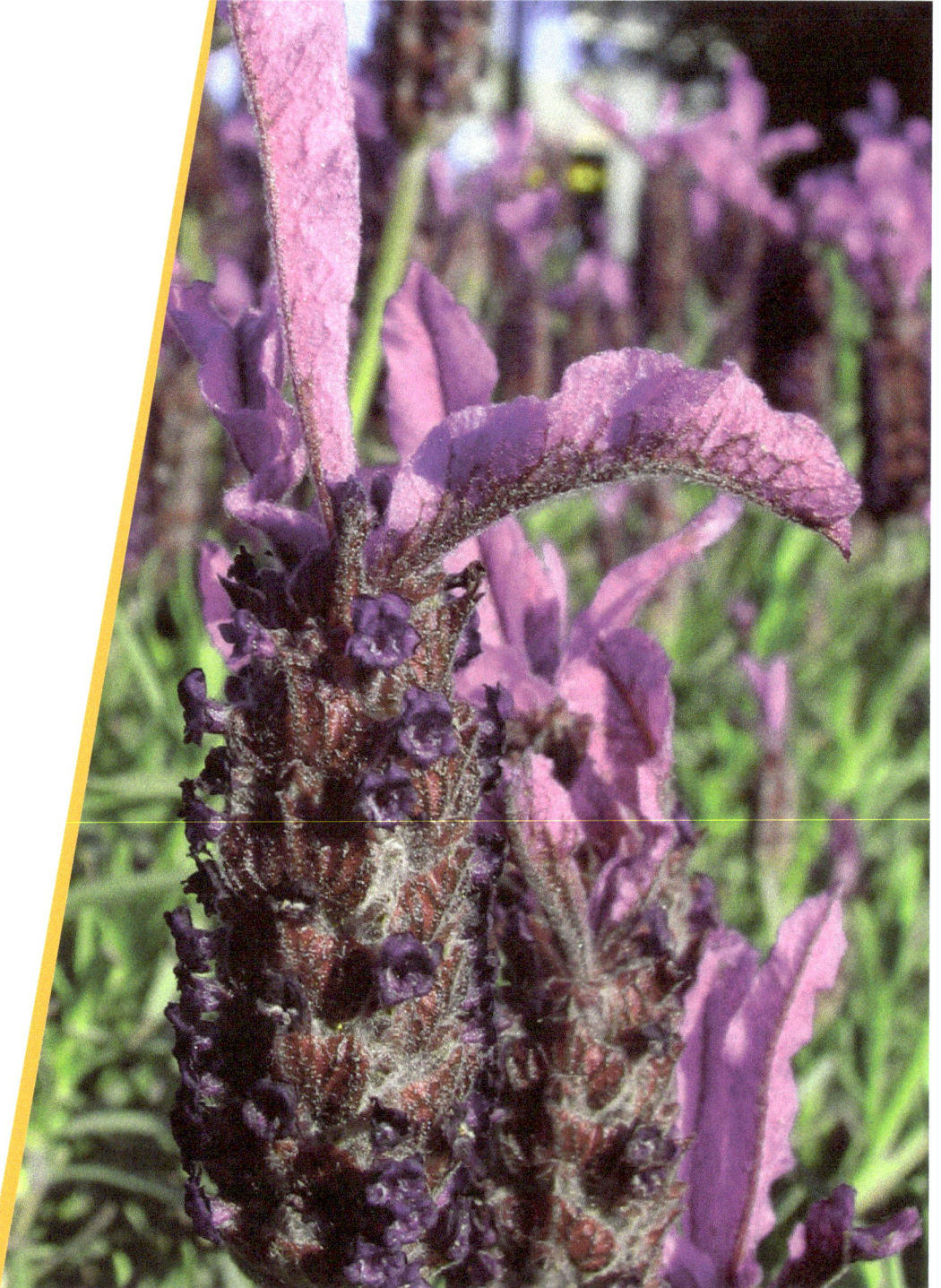

An abbreviated glossary, most terms are explained in context within the book.

| Name | Definition |
| --- | --- |
| abdomen | The soft end or third section of a bee that contains reproductive, digestive and other organs, honey stomach and sting. |
| anaphylactic shock | |
| apiarist | A beekeeper |
| apiary | The location of multiple or even a single hive also called a bee yard. |
| bee bread | A mixture of pollen, nectar or honey, deposited in cells and used to feed young larvae. |
| bee space | A gap of 6-8mm providing room for adult bees to move freely but too small for them to build comb and too big to fill with propolis. |
| beeswax | The building substance of the hive, used for centuries in candles and cosmetics. It is a complex mix of organic compounds secreted by glands on the underside of the worker bees abdomen. Its melting point is 62-64°C |
| brace/burr comb | Brace comb is built to connect between frames or comb and the hive body. Burr comb is often built in the hive lid during a honey flow when there is not enough room. |
| brood | Eggs, larvae and pupae, all the immature stages of the bee. |
| brood chamber | Usually in the lower supers where the brood is reared. |
| capped brood | Laval cells capped with a mix of beeswax and propolis giving a brown colour. Once capped lavae spin their cocoon and turn into pupae. |
| cappings | The thin wax covering on honey cells. |
| clearer board | A board the size of the external dimensions of the hive with one way entries allowing downward movement of bees. Often used to clear bees from |

# GLOSSARY

honey supers to be extracted.

**cell bar**
A wood bar with queen cups attached for queen raising.

**cleansing flight**
Winter and early spring flights in clear weather when bees relieve themselves of waste products.

**cluster**
In winter bees congregate together to conserve heat. Cluster also form as bees cling together during swarming.

**colony**
The single social unit formed by a queen, workers and drones living together in a hive or some suitable cavity.

**comb**
The back to back arrangement of hexagonal wax cells that house, eggs, brood, pollen or honey.

**dances**
Dances are used by returning foragers and scouts to show the location of forage or a new home.

**dance floor**
Where the bee dances occur on brood comb near the entrance.

**division board**
Either the board to divide a hive into two parts or a vertical board used to reduce the size of the brood chamber.

**drifting**
Often juvenile bees will drift to a stronger hive.

**drone comb**
Comb having larger drone cells, about 4 cells per 2.5cm used to raise drones.

**Drone Congregation Area DCA**
A specific area that drones fly to each year in anticipation of mating a virgin queen.

**egg**
The first stage of the life cycle, laid by the queen, a cylindrical form about 1.6mm long.

**extractor**
A machine, manual or electric that extracts honey from comb after the wax cappings have been removed. Often called a spinner as it spins frames and the honey hits the side walls and is collected at the bottom.

**feeder**
Different types of containers used to feed bees sugar syrup.

**field bees or foragers**
Worker bees that are usually 16 days or older that

collect or rob nectar, pollen, propolis, honeydew and water.

**floral sequence**  The calendar of flowering in an area.

**foundation**  The embossed wax embedded in frames or embossed plastic on which bees build comb.

**frame**  A rectangular wooden frame that hangs in a super, contains the foundation/comb. It is made of a top bar, a bottom bar and two end bars and is made from wood or plastic.

**grafting**  Placing newly hatched worker lavae into queen cups to rear queens.

**haemolyph**  The blood like liquid in a bees body. It does not deliver oxygen but provides nutrients and pressure to maintain body shape.

**hamuli**  The fine hooks on the front edge of the rear wing used to lock the front and rear wings together in flight.

**hive**  The home for a bee colony, usually a number of 'supers' rectangular frames carrying brood frames. Honey supers fit on top.

**hive tool**  A tool used to clean up wood ware from wax and propolis and lift frames. Usually made from stainless steel.

**hive Flow®**  An innovation to collect honey directly from the hive.

**hive Langstroth**  Most common modern hive form that incorporates a brood chamber and honey supers. There are a number of configurations using different super sizes. Most common in Tasmania are ideals, manlys and full depth supers.

**hive Top bar**  A simple hive with sloped sides and top bars with starter wax and a brood division.

**hive Warre**  A hive based on a close to nature approach.

**HMF**  Hydroxymethylfurfuraldehyde. A breakdown product of fructose and can indicate overheating. Honey in temperate climates must contain less than 40mg/kg

# GLOSSARY

| | |
|---|---|
| **honey house** | A building used to process honey; extracting, packing and storage. |
| **honey flow or flow** | A time when nectar is flowing freely from flowers of trees shrubs and herbaceous plants. Its honey time. |
| **hypopharyngeal gland** | A pair of glands in a worker bees head that produces brood food and royal jelly. |
| **invertase** | An enzyme bees used to break down the complex sugar in. |
| **LAB Lactic acid bacteria** | A series of bacteria in the honey stomach that prepares pollen and helps prevent fermentation of honey. They may also provide antibiotic like protection for bees. |
| **larva** | The second stage in honeybee (insect development). A white , legless grub like state. |
| **laying worker** | A worker laying unfertilised eggs that turn into drones. This usually from a long period without a queen. |
| **mandibles** | Bee jaws that work in a horizontal motion. Used to construct honey comb, scrape, pickup debris and defend the colony from intruders. |
| **mating flight** | Queens mate in flight and during her first three weeks she must mate with up to 20 drones. |
| **methylglyoxyl** | The chemical compound responsible for antimicrobial activity in Manuka honey from Leptospermum scoparium. |
| **migratory beekeeping** | Moving colonies to follow honey flows from flora in different locations. |
| **Nasanov gland** | Commonly called the scent gland that produces a pheromone that attracts workers. |
| **nectar** | A sugary sweet liquid, a complex sugar, the prime constituent of honey, produced by the nectaries of flowers. |
| **nectaries** | Specialised glands at the base of flowers that produce nectar. |
| **nuc/nucleus hive** | Commonly called a nuc, a small starter hive of 2-5 frames of brood and stores and a new queen. Also |

used for queen rearing.

**pheromone**

Chemical substances used for communication that can change behaviour and physiology of bees.

**pollen**

The critical protein food of bees. It is the male reproductive body produced by anthers of flowers and has a durable casing. Its crude protein content, amino acid composition, digestibility and availability is very important to bee health and colony growth.

**pollen basket or corbicula**

A flattened depression surrounded by curved spines on the bees third leg. Used to carry pollen and propolis back to the hive.

**proboscis**

The sucking tube of the mouth parts used to draw up nectar, honey, honeydew and water.

**propolis**

Bee glue collected from resinous plants/trees used to fill gaps and strengthen comb. It has antimicrobial properties.

**queen**

A fully developed mated female, longer than other bees, usually the mother of all bees in the colony.

**queen cage**

A small cage made from plastic or wire mash used to house a queen and 3-6 worker attendants. Used for queen transport and introduction to a hive.

**queen cell**

A peanut shaped cell about 2.5cm long that hangs vertically within a frame. It can be on the middle of the comb, supersedure, or at the bottom, swarming.

**queen cup**

A cup shaped cell hanging vertically that may become a queen cell. Artificial queen cups are used by breeders on a cell bar.

**queen excluder**

A metal, plastic or wood sheet with gaps only permitting workers to pass. Usually used to restrict the queen to the brood chamber.

**QMP**

Queen Mandibular Pheromone indicates to the colony that the queen is present and active. It is spread by attendants.

**refractometer**

A instrument used to measure moisture content of honey. Stored honey must have 18% or less water to prevent fermentation.

# GLOSSARY

| | |
|---|---|
| **robbing** | Stealing nectar or honey by bees from another colony. |
| **royal jelly** | A highly nutritious substance fed to the queen, queen larvae and all larvae for the first 3 days. |
| **spermatheca** | An organ in the queen where drone sperm is stored and nurtured for the life of the queen. |
| **spiracle** | Breathing tubes along the sides of bees. |
| **split** | Dividing a hive to produce a new colony. |
| **supering** | The addition of supers for the collection of honey. |
| **supersedure** | The natural replacement of an aging queen by a daughter. |
| **swarm** | A group of bees, queen, drones and workers that leave the original colony to form a new one. |
| **thixotropy** | A jelly like state that turns liquid when agitated or stirred. This is common in honey from Leptospermum species. |
| **UMF®** | Unknown Manuka Factor, the original term used for measuring bioactivity in Manuka honey. It is a highly protected trade mark. |
| **Unifloral honey** | The honey derived from predominantly one species. This is checked by measuring different pollen in the honey. |
| **ventriculus** | The stomach in the abdomen of the bee, behind the honey stomach. |
| **virgin queen** | An unmated queen. If she remains unmated is only able to lay unfertilised eggs producing drones. |
| **wax glands** | The eight glands that secrete wax, located in pairs on the last 4 ventral abdominal segments of worker bees. |
| **wax scale** | A drop of liquid wax that hardens to a fish-like scale. These are the building blocks of comb. |
| **winter cluster** | A spherical group of adult bees within the hive during winter. |

| Common Name | Botanical Name | Regions | Frequency (Years) | Flowering Duration (months) | Pollen Quality (1-5) | Honey (kg/hive) | Flowering Time |
|---|---|---|---|---|---|---|---|
| Silver Wattle | *Acacia dealbata* | NW / N / NE / W / SW / SE | 1 | 1-3 | 1-5 | 0-10 | |
| Black Wattle | *Acacia mearnsii* | NW / N / NE / W / SW / SE | 1 | 1-2 | 3-4 | - | |
| Blackwood | *Acacia melanoxylon* | NW / N / NE / W / SW / SE | 1 | 1-2 | 2-5 | - | |
| Boobyalla | *Acacia sophorae* | NW / N / NE / W / SW / SE | 1 | 1-3 | 4-5 | 0-2 | |
| Wattles | *Acacia sp* | NW / N / NE / W / SW / SE | 1 | 2-5 | 2-4 | - | |
| Prickly Mo | *Acacia verticillata* | NW / N / NE / W / SW / SE | 1 | 2 | 3 | - | |
| Waratah | *Agastachys odorata* | NW / N / NE / W / SW / SE | 1 | 1 | 1-3 | 0-1 | |
| Sheoak | *Allocasuarina verticillata* | NW / N / NE / W / SW / SE | 1 | 2 | 3 | - | |
| Horizontal Scrub | *Anodopetalum biglandulosum* | NW / N / NE / W / SW / SE | 1 | 1-2 | 0 | 0-14 | |
| Native Laurel | *Anopterus glanulosus* | NW / N / NE / W / SW / SE | 1 | 1-2 | 0-3 | 0-38 | |
| Golden Pea | *Aotus ericoides* | NW / N / NE / W / SW / SE | 1 | 2 | 4 | 0 | |
| Sassafras | *Atherosperma moschatum* | NW / N / NE / W / SW / SE | 1 | 1-2 | 0-1 | - | |
| Baeura | *Baeura rubioides* | NW / N / NE / W / SW / SE | 1 | 1 | 1 | 0 | |
| Bottle Brush | *Banksia marginata* | NW / N / NE / W / SW / SE | 1-2 | 1-2 | 1-2 | 0-20 | |
| Serrated leaf Banksia | *Banksia serrata* | NW / N / NE / W / SW / SE | 2 | 2 | 1 | 10 | |
| Prickly Box | *Bursaria spinosa* | NW / N / NE / W / SW / SE | 1-7 | 0-4 | 1-5 | 0-50 | |

# NATIVE FLORA

| Common Name | Botanical Name | Regions | Frequency (Years) | Flowering Duration (months) | Pollen Quality (1-5) | Honey (kg/hive) | Flowering Time |
|---|---|---|---|---|---|---|---|
| Yellow Bottlebrush | *Callistemon pallidus* | SE | 1 | 1-2 | 4 | 4-14 | |
| Bottlebrush | *Callistemon viridiflorus* | W | - | 1 | 1 | 0 | |
| Native Plum | *Cenarrhenes nitida* | W, SW | - | - | 2 | 0 | |
| Star Clematis | *Clematis aristata* | SW | 1 | 2 | - | - | |
| Native Hop | *Dodonaea viscosa* | N, SE | 1 | 1-3 | 1-5 | 2-20 | |
| Heaths | *Epacris/various* | NW, N, W, SW, SE | 1 | 1-7 | 1-5 | 0-5 | |
| Unnamed Eucalyptus sp | *Eucalyptus sp* | N, SW | 1 | 1-3 | 1-5 | 2-20 | |
| Black Peppermint | *E. amaygdalina* | N, NE, SE | 0-50 | 1-2 | 1-3 | 0-35 | |
| Snow Gum | *E. cocciferra* | SW | 4 | 2 | 5 | - | |
| White Gum | *E. dalrympleana* | NW, N, NE, SE | 2-10 | 1-2 | 1-5 | 0-30 | |
| White Top | *E. delegatensis* | NW, W, SW | 0-20 | 0-3 | 1-5 | 0-38 | |
| Blue Gum | *E. globulus* | NW, N, NE, SW, SE | 1-12 | 1-7 | 1-5 | 0-60 | |
| Smithton peppermint | *E. nitida* | NW, W | 2-80 | 1-10 | 4-5 | - | |
| Brown Top Stringybark | *E. obliqua* | NW, N, NE, W, SW, SE | 1-15 | 1-4 | 1-5 | 0-38 | |
| Black Gum | *E. ovata* | SE | 1-4 | 1-4 | 3-5 | 8-40 | |
| Cabbage Gum | *E. pauciflora* | SW, SE | 1-4 | 3-4 | 3-5 | - | |
| Peppermint | *E. pulchella* | SE | - | 1 | 4 | - | |
| Swamp Gum | *E. regnans* | SW, SE | 7 | 1-2 | 3 | 0-38 | |

| Common Name | Botanical Name | Regions (active) | Frequency (Years) | Flowering Duration (months) | Pollen Quality (1-5) | Honey (kg/hive) | Flowering Time |
|---|---|---|---|---|---|---|---|
| Ironbark | E. seiberi | NE | 6 | 1-3 | 1 | 30 | Sep–Nov |
| Alpine Yellow Gum | E. subcrenulata | SW | 4 | 1 | 5 | - | Apr |
| White Gum | E. viminalis | NW, N, NE, SE | 2-10 | 1-2 | 1-5 | 0-30 | Jan–Feb; Nov |
| Leatherwood | Eucryphia lucida | NW, W, SW | 1 | 1-5 | 1-5 | 8-115 | Jan–Mar; Dec |
| Deciduous leatherwood | Eucryphia milliganii | SW | 1 | 2 | 1-5 | 0-15 | Jan–Mar |
| Button Grass | Gymnoschoenus sphaerocephalus | W, SW | - | 1-2 | - | - | Jan; Dec |
| Cutting grass | Gahnia grandis | SW | 1 | 1-4 | 1-5 | - | Jan–Mar; Dec |
| Hakea | Hakeasp | N | - | - | - | - | |
| Eastern Berry | Leptecophylla divaricata | N | - | 1-3 | 0-1 | 10-30 | Aug–Sep |
| Mountain Berry | Leptecophylla parvifolia | N | - | 1-3 | 0-1 | 10-30 | Aug–Sep |
| Tea Tree Unnamed | Leptospermum sp | NW, NE, W, SW, SE | 1-5 | 1-12 | 2-5 | 0-40 | Jan–Dec |
| Woolley Tea tree | L. lanigerum | NW, W, SW, SE | 1+-2 | 1-2 | 3-4 | 4-5 | May; Nov–Dec |
| Manuka | L. scoparium | NW, NE, W, SW | 1 | 1-2 | 1-5 | 0-12 | May; Nov–Dec |
| Swamp Paperbark | Melaleuca ericifolia | NW, W | 1 | 2 | 3-5 | 0-13 | May–Jun; Oct–Dec |
| Paperbark | Melaleuca sp | NW, N, SW | 1 | 1-2 | 5 | 0-7 | Nov–Dec |
| Paperbark | M. squarrosa | NE, W | 1 | 2 | 2 | - | May; Nov–Dec |
| Musk | Olearia argophylla | SW | 1 | 2 | 5 | 0-15 | Dec |
| Lancewood | Phebalium squameum | SW | 1-4 | 2 | 4-5 | 5-30 | Dec |

# NATIVE FLORA

| Common Name | Botanical Name | Regions | Frequency (Years) | Flowering Duration (months) | Pollen Quality (1-5) | Honey (kg/hive) | Flowering Time |
|---|---|---|---|---|---|---|---|
| Tallowwood | *Pittosporum bicolor* | NW | 1 | 1-2 | 5 | - | NOV |
| Dogwood | *Pomaderris apetala* | SW | 1 | 2-3 | 5 | 0-14 | JAN; OCT-DEC |
| Christmas Bush | *Prostanthera lasianthos* | SW | 1 | 1-2 | 1-5 | 0-6 | DEC |
| Prickly Beauty | *Pultanaea juniperina* | SW | 1 | 3 | - | - | JAN; NOV-DEC |
| Shaggy Pea | *Pultenaea sp* | NW | 1 | 2-4 | 5 | 0-4 | SEP-NOV |
| Understory | *Various* | NW | 1 | 2 | 5 | - | JAN-FEB; OCT-NOV |
| Wildflower | *Various* | SW | 1-5 | 1-2 | 4-5 | 12-30 | JAN; DEC |
| Yellow Eye | *Xyris operculata* | W | 1 | 1-2 | 1 | - | JAN; DEC |

*Disclaimer: The Floral calendar has been derived from information provided by beekeepers through the 2004 Apairy Industry Census.*

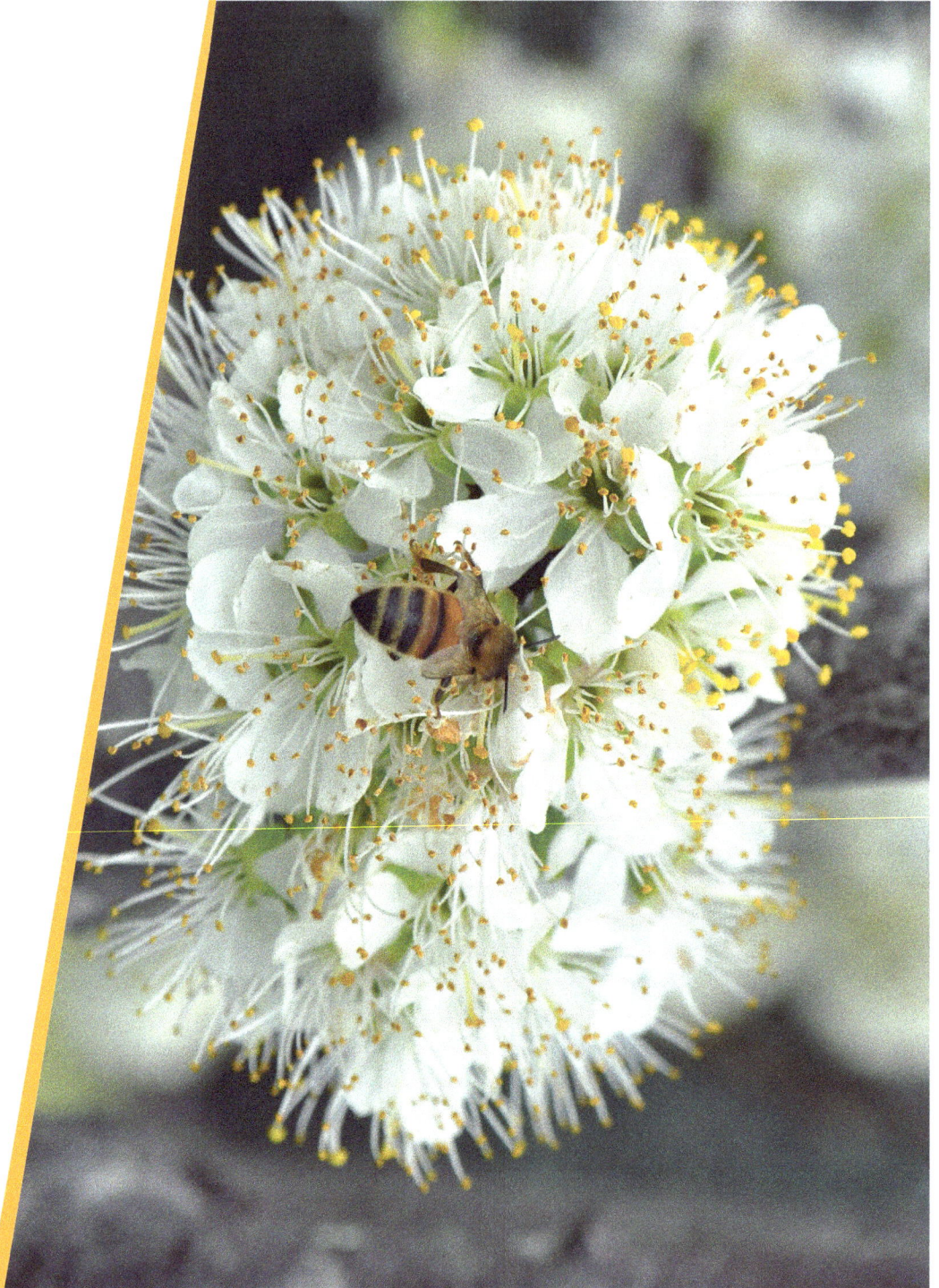

| Common Name | Botanical Name | Regions | Frequency (Years) | Flowering Duration (months) | Pollen Quality (1-5) | Honey (kg/hive) | Flowering Time |
|---|---|---|---|---|---|---|---|
| Sycamore | Acer pseudoplatanum | SE | 1 | 4 | 4 | 2 | |
| Garlic | Allium sativum | SE | 1 | 4 | 4 | 0 | |
| Cape Weed (w) | Arctotheca calendula | NW, N, NE, SE | 1-4 | 1-4 | 2-5 | 0-60 | |
| Mustard | Brassica juncea | SE | 1 | 3-4 | 4 | 0 | |
| Cabbage | Brassica olearea capittat alba | SE | 1 | 1-2 | 4 | 0 | |
| Canola | Brassica rapa | SE | 1 | 2 | 4-5 | 0 | |
| Wild turnip (w) | Brassica rapa ssp campestris | NW, N, SE | 1 | 1-3 | 1-5 | 0 | |
| Catoneaster | Catoneaster sp | SE | 1 | 1 | 4 | 5 | |
| Blue Pacific | Ceanothus sp | NW, N, NE, W, SW, SE | 1 | - | - | - | |
| Hawthorn (w) | Cystus palmensis | NW, N, NE, SE | 1 | 1 | 2-5 | 10 | |
| Yellow Tree Lucerne | Crataegus monogyna | SE | 1 | 2 | 5 | 1 | |
| Broome (w) | Cystus scoparius | N | 1 | 1 | 5 | 20 | |
| Pride of Madera | Echium fastuosum | NW, N, NE, W, SW, SE | 1 | - | - | - | |
| Salvation Jane (w) | Echium plantagineum | SE | 1 | 3 | 4 | 0 | |
| Spanish Heath (w) | Erica lusitanica | NW | 1 | 2 | 5 | 0 | |
| Lavender | Lavandula sp | NW, N, NE, W, SW, SE | 1 | - | - | 30 | |

# NON-NATIVE FLORA

| Common Name | Botanical Name | Regions | Frequency (Years) | Flowering Duration (months) | Pollen Quality (1-5) | Honey (kg/hive) | Flowering Time |
|---|---|---|---|---|---|---|---|
| Lotus major | Lotus major | NE | 1 | - | 3 | - | |
| Apples | Malus sp | N, SE | 1 | 1 | - | - | OCT–NOV |
| Lucern | Medicago sativa | SE | 1 | 1-2 | 1-5 | 4-10 | JAN; OCT–DEC |
| Scotch Thistle (w) | Onopordium acantheum | SE | 1 | 2 | 2 | 0 | SEP–OCT |
| Opium Poppies | Papaver somniferum | NW, SE | 1 | 1 | 4 | 10 | NOV–DEC |
| Cherries | Prunus | NW, N, NE, W, SW, SE | 1 | 1 | 5 | 0 | OCT–NOV |
| Apricots | Prunus armeniaca | N, SE | 1 | 1 | 5 | 0 | SEP |
| Wild Raddish (w) | Raphanus raphanistrum | SE | 1 | 3 | 3 | 0-5 | OCT |
| Blackberry (w) | Rubus fruticosus | NW, N, NE, SE | 1 | 1-3 | 3-5 | 0-60 | JAN–FEB; OCT–DEC |
| Currants | Rubus nigrum | SE | 1 | 1-4 | 2-5 | 0-5 | OCT |
| Berries | Rubus sp | SE | 1 | 3 | - | - | NOV–DEC |
| Loganberries | Rubus sp | SE | 1 | 4 | 2 | - | JAN; OCT–NOV |
| Raspberries | Rubus sp | SE | 1 | 2 | - | - | JAN; DEC |
| Weeping Willow | Salix babylonica | SE | 1 | 4 | 2 | - | SEP–OCT |
| Pussy Willow | Salix discolor | SE | 1 | 4 | 2 | - | SEP–OCT |
| Crack Willow (w) | Salix fragilis | N, NE, SE | 1-3 | 4-5 | 0-8 | | SEP; SEP–OCT |
| Ragwort (w) | Senecia jacobaea | NW, N, NE, W, SW, SE | - | - | - | - | JAN–FEB; OCT–NOV |
| Charlock (w) | Sinapsis arvensis | SE | 2 | 1 | 3 | 1 | OCT |

# NON-NATIVE FLORA

| Common Name | Botanical Name | Regions | Frequency (Years) | Flowering Duration (months) | Pollen Quality (1-5) | Honey (kg/hive) | Flowering Time |
|---|---|---|---|---|---|---|---|
| Potato | Solanum tuberosum | NW N NE W SW SE | 1 | - | - | - | |
| Pyrethrum | Tanacetum cinerariifolium | NW N NE W SW SE | 1 | - | - | - | |
| Dandelion (w) | Taraxacum officinale | NW N NE W SW SE | 1 | 1-5 | 1-5 | 0-16 | |
| Linden | Tilia europaea | NW N NE W SW SE | 1 | 1-2 | 2 | 0 | |
| Clover (various) | Trifolium sp | NW N NE W SW SE | 1-15 | 1-4 | 1-5 | 0-37 | |
| Gorse (w) | Ulex europaea | NW N NE W SW SE | 1 | 1-6 | 4-5 | 0-2 | |
| Garden Flora | Various | NW N NE W SW SE | 1 | 1-12 | 2-5 | 0-30 | |
| Home Orchard | Various | NW N NE W SW SE | 1 | 2-4 | 4 | - | |
| Grapes | Vitis vinifera | NW N NE W SW SE | 1 | 1 | 1 | 4 | |

*Disclaimer: The Floral calendar has been derived from information provided by beekeepers through the 2004 Apairy Industry Census.*

*Wineglass Bay Freycinet Peninsula. Image credit Paul Morton/iStock by Getty Images*

*Wineglass Bay Freycinet Peninsula. Image credit Paul Morton/iStock by Getty Images*